RECYCLE WITH EARTHWORMS

The Red Wiggler Connection

Shelley Grossman
&
Melissa (Toby) Weitzel

Editor: Lucy Warren
Illustrations: Lisa Marie Donnabella

SHIELDS PUBLICATIONS

PO Box 669   Eagle River, WI 54521

Library of Congress Control Number: 38889546

ISBN 13: 978-0-914116-32-5
ISBN 10: 0-914116-32-0

Dedication

We dedicate this book to all of you who share our love of the earth and, of course our hero and star, the earthworm.

Acknowledgments

We were humbled by the willingness of very busy people who gave of their precious time, expertise and knowledge to assist us. We needed guidance, wisdom, facts and technical support to write this book. We found those qualities and attributes in the people and organizations listed below. For that unstinting help, we say thank you.

Stephan White, Associate Editor, Worm Digest; Clive Edwards of C.A. Edwards and P.J. Bohlen, authors of *Biology and Ecology of Earthworns*; Mary Appelhoff, author of *Worms Eat My Garbage* and *Worms Eat Our Garbage*; Uday Bhawalker of the Bhawalker Earthworm Institute, Pune, India; Vincent Lazaneo, Director Home Farm Advisor at the University of California Cooperative Extension; Lisa Marie Donnabella, our friend and graphic artist; Ralph Laughlin, Master Composter and Master Gardener; Doug Hoblit, Master Composter and Vermiculturalist; and the Plastic Bag Information Clearing House of America. For the inspiration imparted to us by the legions of authors of children's worm books and by the entire group at the California Integrated Waste Management Board with special appreciation to Nancy Carr.

In our local community of Carlsbad, California: Dr. Susan Bentley, Superintendent of Instructional Services, Carlsbad Elementary School District; the Carlsbad School Board of Trustees, 1995 through 1997; Steve Ahle, Principal, Pine and Jefferson Elementary Schools; The Honorable Bud Lewis, Mayor, and City Council Members, 1995 through 1997; Carlsbad Farmer's Market; Carlsbad Parks and Recreation Department.

Lucy Warren, our editor, mentor and friend, for her keen insight, humor and unfailing patience, our heartfelt thanks. We couldn't have done the book without you.

To our husbands and children: thank you for believing in our dream.

Table of Contents

Introduction - The Importance of Composting   1

Beneath it All - Soil   5

Effects of Earthworms in Soil   8

About Earthworms   12

Composting Masters - Red Wigglers   16

Worm Anatomy and Physiology   19

Worm Reproduction and Development   27

Keeping Worms   32

Setting Up a Worm Bin   34

Feeding Earthworms - Do's and Don'ts   43

Things Happen   48

Worm Bin Trouble Shooting Guide   56

Special Situations   58

Harvesting Worm Castings   60

Worm Myths   69

Appendix   73

     Plastic vs. Paper in the Waste Stream   75

     Common Organic Waste Resources   77

     Information and Resources for
        Worm Related Items   79

     Glossary   81

     Further Reading   87

Index   90

Introduction - The Importance of Composting

We live in a "throw away" society. As mankind rushes pell mell into the technological and information ages many of us have lost our place in nature. Farm conglomerates strip our farmlands of valuable nutrients, adding back chemical forms of fertilizers. Corporations log our forests for increasing demands of lumber and paper. Homeowners landscape our homes with exotic plants and beautiful lawns which require intensive care.

The populace must then contend with the byproducts of this "progress" — the generation of literally tons of waste materials, in need of disposal. This trash fills up enormous landfills at staggering rates. Based on the national average of 3.2 persons per household, each household generates approximately one ton of solid waste per year or 600 pounds per person.

In the past our resources seemed unlimited. We would simply send this material off to immense landfills where it would be buried and forgotten. But each year the waste grows and the space for landfills diminishes. Learning more, we become concerned that the byproducts of that waste may leach into our water supply or cause other harm. And, costs continue to rise. The resources being sent to landfills become entombed, not recycled for continued benefit, in a costly tax supported institution. Fortunately there are alternatives.

The waste consists of both organic and inorganic matter. Organic matter is anything which is living or has ever lived, plant or animal. Organic matter includes paper, yard trimmings, food stuffs, some clothing, petroleum and wood products, to name a few examples. Organic waste can be broken down into simpler compounds – or composted – to create a natural soil amendment with nutrients readily usable by plants. Inorganic matter is composed of minerals, anything which is not made of living or once living animals or plants.

In California, organic waste makes up 30-40% of the waste sent to landfills. The estimated figure for kitchen waste is a little over 10% of all food brought into the household![1] By diverting kitchen waste, we can save landfill space and replenish our soils. Many states, cities and counties have already put regulations in place, mandating a decrease in the volume of materials sent to landfills. This means that if citizens do not comply, they could be fined. In California the target date for fifty percent waste stream reduction is the year 2000.

By diverting all or part of your kitchen, paper and yard waste you will save landfill space and gain a valuable soil amendment. Recycling, composting and vermicomposting are the logical steps to aid in our personal, local, national, even global ability to deal with the growing trash crisis.

Moreover, much of the now wasted material can readily be put to better use. A good example is the daily dilemma of paper versus plastic bags at the grocery store. Do you take paper, knowing a tree has been sacrificed; or plastic, knowing it won't breakdown in the landfill? Fortunately many stores have bins to receive plastic bags and other clear plastics for recycling.[2]

While the problem is staggering, the impact of individuals simply taking responsibility for their own waste will go a long way toward a solution. Just remember the entire ocean is made up of little drops of water. Our landfills are made up of little and big pieces of trash so every effort, even a seemingly small one, helps.

---

[1] Based on data provided by the California Solid Waste Management Board

[2] See Appendix for further information.

Random City Saves Tons of Trash and Greens Up With Worms

Using worms to compost kitchen waste can have a big impact on the waste stream.

Random City in the state of Fiction, USA has 5,000 households. Each household creates four pounds of kitchen waste per week.

In response to increasing landfill costs and ecological concerns, the voters decided that henceforth all kitchen waste will be fed to worms. Let's look at the impact:

5000 houses x 4 lbs. waste per week = 20,000 lbs.
one ton = 2,000 pounds
20,000 lbs ÷ 2,000 lbs/ton = 10 tons
10 tons/week x 52 weeks/year = 520 tons waste/year

At the end of the year Random City residents rejoiced because they not only saved their landfill fees but also had healthier plants and greener gardens by using their worm castings!

Gardening has become a thriving hobby of Random citizens. They often compare notes on the health and well being of their carefully tended worms.

Composting and vermicomposting (composting with worms) are logical steps to deal with the growing trash/waste crisis and diminishing soil fertility. From the beginning of time organic waste has been recycled through natural decomposition. Dead matter falls to the earth to be consumed by numerous organisms making the nutrients available to future generations of living plants and animals. This process returns nutrients to the earth and helps build dynamic, fertile soil.

In the past few centuries humans have created ways to alter or halt Nature's balanced system. By sealing our organic waste in landfills and only using commercial chemical soil amendments, mankind has deprived the soil of rich organic and inorganic elements it needs to remain a sustainable resource and to produce healthy crops. While the chemical soil amendments and fertilizers provide the proper nutrients, they don't add back the essential organic soil components, humus and microorganisms, which facilitate good plant growth. But it's not too late! Through composting and vermicomposting we can replenish those primary elements essential to healthy plant growth.

In the following sections you will learn about soils and soil structure, worms and their physiology, and how to construct a vermicomposting system to process household and kitchen waste while creating a rich organic soil amendment.

Please note that although we offer a lot of explanation in order to be thorough and to assure your success, the process is quite easy and time efficient.

Beneath it All - Soil

Soil is defined variously as: "the loose upper layer of earth in which plants grow," from the *Oxford American Dictionary*; "the portion of the earth's surface consisting of disintegrated rock and humus," *Webster's Encyclopedic Unabridged Dictionary.* The critical thing to know is that soil has two parts, the organic humus and the inorganic base of decomposed rocks.

We usually take it for granted but the soil in which our food crops grow determines their ability to sustain life. A well balanced, nutrient rich soil with good porosity and ample water will produce healthier, stronger, more disease resistant crops. In turn, those crops provide the building blocks for the food chain including ourselves.

Inorganic Base

One basic ingredient in soil is ground up rock, the stuff that makes up the earth. Soils are defined by the particle size of this rock. The particles can be large or small and are categorized as sand, silt or clay according to their size, largest to smallest respectively. Most soils have a variety of particle sizes but the one in greatest abundance by volume takes precedence in its description. An example would be: silty clay, more clay than silt.

Sand feels gritty and has the largest particle size. The particles stay separate and don't hold together, making it very porous to both water and air. Sandy soil is easy to cultivate. The relatively large spaces between particles allow water and air to pass through quickly and easily. Unfortunately, sandy soils do not retain very much of the water or minerals to promote plant growth.

Silt feels smooth when dry, and slippery (but not sticky) when wet. As silt is made up of the intermediate size particles, it has smaller pore spaces than sand and absorbs water less quickly but holds it better. It also has low fertility and does not retain minerals well. At the other extreme, clay has the smallest particles. The particles fit so closely together that it is difficult for water, air and roots to penetrate. However, once absorbed, clay has an affinity for and can retain water to the point of becoming water logged. When wet, clay soil is sticky to the touch. Clay soil retains nutrients well, its chemical properties attract and hold onto them. However, because clay soil compacts easily, plant roots have difficulty reaching the rich store of nutrients.

Soil good for growing plants allows water to percolate through to the plant root zone while retaining some of the moisture on the soil particles for future use, but not so much as to drown the plant. Good soil is porous enough so that air can also reach the roots of the plants. It has an abundance of balanced nutrients in forms readily available to the plants.

Loam soils have combinations of all three particles sizes and are the most versatile for growing healthy plants. Fortunately it is possible to modify basic soil characteristics by adjusting their organic component.

Soil Organics

The second component of soil is organic matter called humus. *Webster's Encyclopedic Unabridged Dictionary* defines humus as "the dark organic material in soils, produced by the decomposition of vegetable or animal matter and essential to the fertility of the earth." Organic matter – humus – not only provides plants with nutrition in a form which they can readily absorb, but also the decomposed organic matter greatly affects the structure of the soil.

Just as sand, silt and clay are the result of inorganic decomposition of rocks, humic acid is a product of organic decomposition. It is a sticky substance which helps to bind soil particles together in groups, called aggregates.

Aggregated soils are less susceptible to compacting because the spaces between the tiny clumps of soil particles are more porous. These tiny clumps of particles better hold nutrients and water which are more easily absorbed and retained by the humic acid. This makes these soils more amenable to air, water and root penetration; and better able to optimize water retention than unaggregated soils. Blended soils with ample amounts of humic acid are easy to dig and enable plants to grow more easily.

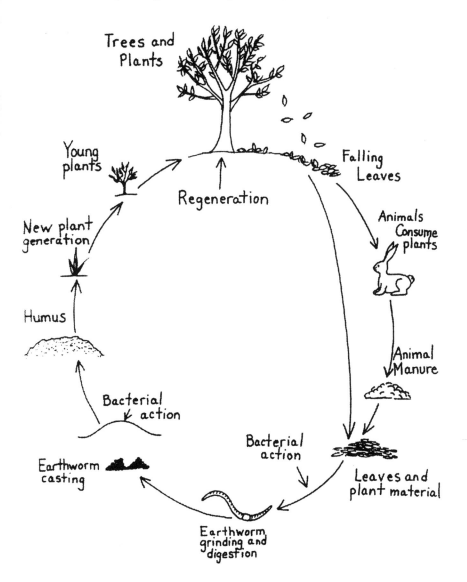

Effects of Earthworms in Soil

Worms are nature's soil processors and mix masters. Worms eat and digest organic matter. They also burrow their way up, down, around and through the soil, mixing up the mineral soil particles of all sizes with the organics. Worm burrows create air passages necessary for good root development and provide channels that ease root hair growth. Thus worms enhance soil porosity for plants.

But worms help with more than just porosity. Over the ages, rich, fertile soils have passed through worms countless times recycling plant material as it decomposes, in turn providing an enhanced nutrient source for plants. Worm manure, also called castings, is a rich, readily absorbable source of nutrients for plants.

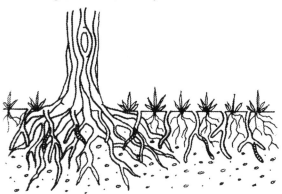

Worms play a critical role in soil reclamation and sustained plant health. By processing the organic waste into the soil, they counter the effects of poor land manage-ment, wind, water and rain erosion, thereby bringing soil back to a more natural balance. Thus encouraging plant growth, worm activity reduces the devastating effects of all types of erosion and helps soils to retain life sustaining nutrients. But those organics (decaying plant matter) have to be available for the worms to do their work!

The worms digest organic matter and excrete it in the form of castings or vermicompost. This black, granular, humus-rich material can be reapplied to enhance your soil. The castings are an effective slow-release organic fertilizer. We can't guarantee the nutrient composition of worm castings exactly the way it can be done with chemical salts. If one were to see it listed like fertilizers, it would contain the following ranges of principal nutrients:

N (Nitrogen)	1.5-2.2%
P (Phosphorous)	1.8-2.2%
K (Potassium)	1.0-1.5%

Variations are due to the worms' diets.

The castings slowly release nutrients over several years. They can be used anywhere because castings will not burn plants. When produced in a home worm bin, they are a no-cost alternative to chemical fertilizers.

Worm processing adds value. Commercially, worm compost or vermicompost sells for up to twice as much as regular compost and has a good reputation with growers for healthy plant production. By adding a worm bin to your composting efforts, you are increasing your soil and plant health.

As worms move through the soil, eating and excreting, something wonderful happens. Within the gut of the worm the soil and organic matter are mixed. The worm uses decayed organic matter as food to provide its own energy and sustenance. This food is conditioned by saliva in the worm's mouth. As the food particles make their way through the worm's intestine they are combined with microscopic bacteria, fungi, mold and actinomycetes.

Finally, the worms excrete nutrient-rich organic byproducts called manure or castings, which can be taken up and used by plants. The microorganisms excreted with the decayed matter and mineral particles further contribute to soil health, creating a rich soil teeming with life.

Upon excretion the processed food is encased in a mucus sheath. This sheath acts as a binder, holding the processed particles and surrounding soil particles together. It dissolves gradually making the nutrient-rich worm castings nature's time-release food for plants.

In *The Biology and Ecology of Earthworms,* by C. A. Edwards and P. J. Bohlen the authors cite studies showing a great increase in the nutrient value of castings compared to adjacent soil.[3]

The same source depicts additional studies showing dramatic increase in fine hair root growth in the presence of worm burrows and castings. Root hairs absorb nutrients from the soil. Worm castings provide plant sustenance in a form readily available for absorption. Roots also take advantage of worm burrows as growing space, filling in the tunnels the worms leave behind.[4]

Other than limited ventures in hydroponics, we grow our food in soil, and soil rich in nutrients is essential for our survival. Worms play a critical role in making the minerals in the soil available to plants. Those plants, in turn, provide for our own nutritional needs.

By composting organic wastes, we divert organics from the overburdened trash waste system and produce a nutrient-rich soil amendment. Even better, it's free!

---

[3] Page 224.

[4] *Ibid.* page 227.

Consider the immense impact you can make in your home, town, and country when you begin to view garbage (food, paper, grass) as a free organic resource instead of "waste." This is the first step in restoring the balance of nature. Worms are a wonderful vehicle for recycling our organic excess.

About Earthworms

Are all worms the same?

Earthworms are a part of the phylum *Annelida* which includes over 9,000 species;[5] 3,000 plus are worms. They are called *annelids* because they have no skeletons and are segmented into rings. They are as diverse in appearance, habitat, mating practices and diet as some of earth's other large animal groups. For instance, birds also vary in size, color, habitat, and diet, but they are all still recognized as birds. So it is with worms.

There are giant worms and minuscule worms. The *Megascolides australis* is a burrowing worm in Australia. Reddish brown, it grows to six feet in length and weights up to twelve pounds. In contrast, the *Enchytraeidae*, an aquatic worm, is white, one-half inch long and hair thin.

Some worm species are single sexed, male or female, however, most are hermaphroditic having sex organs for both male and female in one worm. But even among the hermaphrodites, there are reproductive differences.

Worms can be white, green, brown, red, blueish and many hues in-between. Their dietary preferences vary by species and habitat.

Worms exist in nearly all the earth's environments. They can be found in our lakes, ponds, oceans, forests, plains and mountains...nearly anywhere sufficient decaying organic matter can be found. However, they can't survive volcanic flows, deep desert or permafrost at sub-zero temperatures.

One useful categorization of worms is by their habitat. Aquatic worms are water dwellers, subaquatic live partly in water and partly on land, and terrestrial live only in soil which is why they are also commonly called earthworms.

---

[5]Grzimek's *Animal Life Encyclopedia*, Volume 1, "Lower Animals", p. 362.

Soil bound worms are recognized by the soil horizon level in which they are commonly found, that is, leaf litter, topsoil, and subsoil. There are three major ecological groups as follow:
1) *Epigeic* worms live on the soil surface and uppermost mineral soil. In Latin *epi* means "on" or "upon" and *geic* means "earth." These worms reproduce at relatively high rates and grow rapidly. They form no permanent burrows.

2) *Anecic* worms fall into the same category as topsoil species but form semi-permanent, vertical burrows in the soil which descend from the surface to the mineral horizon. Thus they inhabit the layer between the rich organic surface and the mineral base. They create burrows as they search for and process food.

3) *Endogeic* worms inhabit the deeper mineral horizons. They create permanent horizontal burrows. They consume more soil than other worms, reproduce more slowly and in fewer numbers. In Latin *endo* means "within."

These three groups, working together, burrow, mix, amend and develop top soil.

The most common worms in North America are the burrowers and top feeders, of which there are many species. It would take a lifetime to study worms in depth, so we will concentrate on the species of burrowers and top feeders you are most likely to encounter in your gardening efforts.

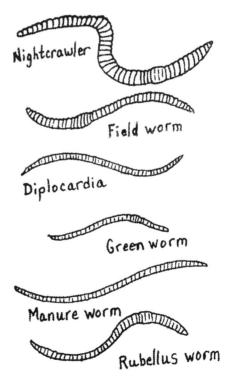

Nightcrawler

Field worm

Diplocardia

Green worm

Manure worm

Rubellus worm

Burrowers

Burrowing worms are the primary agents responsible for mixing soil particle types. One of the most common burrowers in North America is *Lumbricus terrestris*. This burrower is robust, almost pencil thick, and frequently grows up to six inches long. It lives in soil depths from six inches to six feet.

By the very action of burrowing through the soil, *Lumbricus terrestris* creates tunnels that allow air and water to reach down into plant root zones. Burrowers subsequently use their tunnels as underground throughways to move about, even creating chambers for winter hibernation. At the depths where they live the ground is stable and the burrows can last over time.

Burrowers move through the soil eating organic matter, dirt . . .whatever is before them. These worms create tunnels that allow water, air, and nutrients to filter down to the root level of plants, trees and shrubs.

Another common burrowing worm *Lumbricus rubellus* has a distinctive red color and is often called a red worm. This points out the difficulty in using common names for specific animals, as the term "red worm" is used to describe at least two different worms with reddish coloration.

Top Feeders

The top feeders have a slightly different function. Living in the top six to twelve inches of the soil, they devour the large amounts of decaying organic matter that fall to the earth's surface. In the wild these voracious eaters consume up to their own weight in organic matter each day. They leave behind their nutrient rich manure, referred to variously as casts, worm castings, worm compost, or vermicompost. Vermicompost is an excellent source of nutrients for plants, an organic and dynamic soil amendment.

One of the most common top feeders in the United States is *Eisenia fetida*, highly recommended for vermicomposting. *E. fetida* is another worm referred to as a red worm because of its color.

Top feeders differ from burrowers in that they do not create permanent tunnels. The material where they live is loose and whatever space they have created by moving through the area is soon compacted. They leave tunnels behind, but only as a consequence of their movement. These are not used as a permanent throughway and are crushed or destroyed as soon as a human or animal walks above them.

Composting Masters - Red Wigglers

From this point we focus on the top feeder *Eisenia fetida*, the skunk of the worm family! *Fetida* is Latin for "to stink." The reference comes from coelomic fluid inside the worm body, which is foul tasting to predators. Like the skunk, this foul emission is the worm's means of self defense. Also like the skunk, the yellow fluid is squirted out behind the worm when it is startled or in danger. Coelomic fluid is harmless to us, but distasteful to a bird or other predator attempting to devour our squiggly friend.

Eisenia fetida are commonly known as red wigglers or composting worms. They live just below the soil surface, in the layers of rotting organic material. In the wild these worms are sometimes called: top feeders, bandlings, or manure worms.

E. fetida is a hardy breed, readily adaptable to captivity, dietary changes and more tolerant of temperature extremes. They eat more, reproduce faster, and in greater numbers than the burrowers. For these reasons, we recommend using *E. fetida* for your vermicomposting efforts.

Will red wigglers eat my plants?

Worms do not damage roots or foliage of the plants in your garden. Their role comes fairly late in the decomposition cycle. They are always in search of organic matter that has been partially broken down by bacterial decomposition. They are, in essence, the clean up crew.

They have microscopic mouths, no teeth and are dependent on microbes and other decomposing organisms to break down the food for their consumption. Once found, the food is moved to the mouth's opening by the *prostomium*. This fleshy nub acts as an appendage, functioning much like the tip of an elephant's trunk. Instead of chewing, worms act like so many tiny vacuum cleaners.

All digestion of worm food goes on in the digestive system and does not start in the mouth as in humans.

Where do red wigglers live?
In the wild *Eisenia fetida* live on the soil surface, just under the leaf litter. The decaying organic matter on top of the soil supplies worms' dietary needs.

Why do worms need moisture?
The best living space for red wigglers is moist and out of the sun. When exposed to direct sunlight for *three minutes or less*, the worms will rapidly dehydrate and die. Worms breathe through their skin, so a moist environment helps them to breathe. Moisture helps worms to move around their environment more easily. Locomotion is necessary for them to feed and mate.

If earthworms need so much moisture, can they swim?
No, they can't, even though we frequently see them in puddles after a good rain. Worms can survive in water. While the air we breathe is 21% oxygen, the oxygen levels in water are far lower. Occasionally terrestrial worms adapt to living in water and turn a lovely burgundy color, but they really prefer to live in their natural habitat near the surface of the earth.

Do I have composting worms in my soil?
The following simple test will help you check for top feeders in your soil. Soak a piece corrugated cardboard overnight. At dusk the next day, place the cardboard on the top of the soil surface. In the morning, carefully pull the cardboard apart.

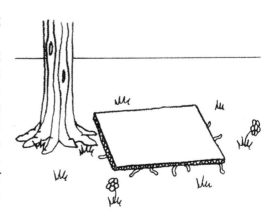

If you have top feeders in your soil you will find them between the corrugations. You may also see some parts of the cardboard missing, a snack for hungry worms.

Worm Facts — Things to Know

- Worms are animals classified as annelids.
- Worms are nocturnal.
- Direct exposure to sunlight can be fatal in less than three minutes.
- Worms breathe through their skin.
- The first one-third of the body contains the vital organs.
- The remaining two-thirds contains the intestine.
- In captivity, a worm will consume one-half its weight daily.
- One-quarter pound worms per square foot surface area is ideal to begin a worm bin.
- At about ¾ pound per square foot worms get overcrowded.
- Like humans, worms do well with a varied diet.
- Calcium is necessary for worm reproduction.
- A neutral environment is necessary with pH greater than 5 but less than 9.
- Ideal temperature range: 55 - 85°F (13 - 29 °C).
- Salt is harmful to worms and should only be given in small quantities.
- Soak dry foods and salty foods in water before feeding to worms.
- Dampness is essential, bin contents must have moisture saturation of 80-90%.

Worm Anatomy and Physiology

Even though they don't have a skeleton and can't walk upright, worms do many of the same things people do to survive. They move around, eat, breathe, reproduce and defend themselves. They are sensitive to temperature, moisture, light and vibrations. In learning about the anatomy of the worm, we hope you will be more sensitive to their needs and better prepared to care for them.

What are annelids?

Worms are annelids, from the Latin word *anulus* meaning "ring." Worms are made up of joined, ringed segments. An adult *Eisenia fetida* has between 80 to 120 circular rings. Try thinking of the giant sequoia trees and their rings. The tree rings grow around each other while the worm rings are stacked end on end.

How is the body constructed?

The cuticle is the worm's outermost body wall. Beneath the cuticle are:
1) the epidermis, which is like our skin,
2) a layer of nerve tissue which performs like our sense of touch,
3) circular and longitudinal muscles for locomotion.

The epidermis contains many sensory cells that transmit information to the nerve tissue. Within the layer of nerve tissue are cells that forward sensory information to the worm's nerve cord and on to the cerebral ganglion, the worm's version of a brain.

Circular muscles create the worm's body rings. These muscles contract and expand, shortening and lengthening the worm's body. The longitudinal muscles run the length of the worm. Acting in concert these sets of muscles enable the worm to propel forward, backward and sideways. Moisture in their environment lubricates this locomotion.

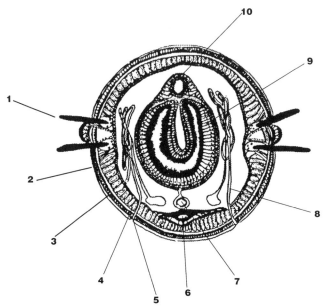

1 Setae
2 Epidermis
3 Circular Muscles
4 Transverse Muscles
5 Excretory Pore
6 Ventral Nerve Cord
7 Subneural Blood Vessel
8 Nephridium
9 Muscular Wall of Intestine
10 Dorsal Blood Vessel

Additionally, stiff hair-like protrusions called *setae* stick out of almost every ring along each side of the worm. These protrusions grip the soil, bedding, or any material it is moving through so the worm doesn't just slip and slide randomly. The setae are made of *chitin*, the same substance that makes up our fingernails and the exoskeleton of many insects. The setae are very strong, assisting the worm as it moves through its environment. The setae help the worm to defend itself by gripping the soil when attacked by a hungry bird or other predator.

The Brain and Nervous System

The cerebral ganglion, located at the front of the worm, serves as the brain. This nerve bundle is responsible for receiving external information such as light, heat, moisture and vibrations. The worm relies on the ganglion and a ventral nerve cord for sensory input from the world around them.

While we don't fully understand all the functions of the nervous system, it is believed that body functions such as reproduction and life cycles are regulated within the nerve ganglion.

Circulation

The circulatory system is powered by five pseudo-hearts. These hearts are merely valved chambers that regulate blood flow and produce a pulse. Branching off these hearts are both a dorsal (forward flow) and a ventral (backward flow) blood vessel. The dorsal and ventral vessels transport the blood, rich with oxygen and nutrients, through the body. The circulatory system also transports urinary waste, which is diffused through the cuticle, the outer covering, in each ringed segment. In other words, worms breathe and excrete urine through their skin.

INTERNAL ANATOMY

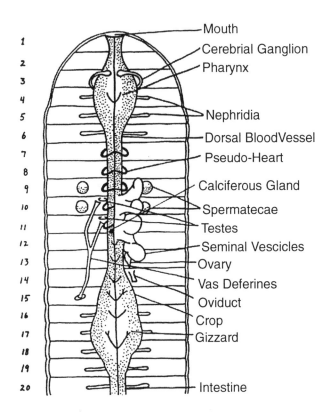

1
2
3
4
5
6
7
8
9
10
11
12
13
14
15
16
17
18
19
20

Mouth
Cerebrial Ganglion
Pharynx

Nephridia
Dorsal BloodVessel
Pseudo-Heart
Calciferous Gland
Spermatecae
Testes
Seminal Vescicles
Ovary
Vas Deferines
Oviduct
Crop
Gizzard

Intestine

Digestion/Gastrointestinal System

Running through the worm's body is the alimentary canal or gut. It starts at the mouth, called the *buccal cavity*, and moves to the back with the pharynx, esophagus, crop, gizzard, intestine, and anus, respectively.

The buccal cavity contains specialized sensory cells which allow worms to locate food and minerals. The cells detect and recognize sucrose, glucose, quinine and saline chemicals from the environment. This allows them to identify and select the foods they eat.

The pharynx works like a suction pump, drawing particles farther in from the mouth. The esophagus, which opens from the pharynx as a narrow tube, leads to the crop. Worms and birds both use their crop as a food storage chamber.

Next is the gizzard, the food grinding chamber. It contains sandy grit from the soil to pulverize food into small particles, including leaf litter, mulch and soil organics.

The intestine is a tube going straight back to the anus, taking up almost two-thirds of the length of the worm. The intestine performs the final digestion and absorption of the life sustaining nutrients from the worm's food.

Enzyme Benefits

Many tiny organisms, bacteria, fungi, actinomycetes, enzymes and protozoa live in the worm's gastrointestinal systems aiding digestion. They are microscopic and thrive by the hundreds of thousands within a single worm. These organisms assist in preparing the nutrients to be absorbed and utilized by the worm.

The worm produces numerous enzymes which aid in its own survival, including an insecticide and an antibiotic.

These enzymes emulsify with mucus produced in the worm's gut and sheathe the castings when expelled through the anus. Plants are able to absorb the insecticidal and antibiotic enzymes through their roots to further utilize them in the plants' ongoing battle to ward off insects and disease. The antibiotic enzymes

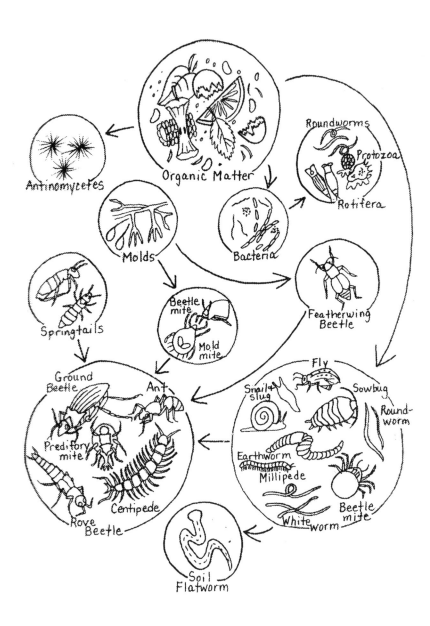

also protect humans from harmful bacteria while we are working in the bin.[6]

How do Worms Really Eat?

Let's say banana peels are on the worms' menu today and into the bin they go. Back to the house we skip, happily knowing we have provided a yummy feast for our worms.

Two days later we are out to check the bin, food bucket in hand, and what do we see as soon as we lift the lid? Our banana peel, a little moldy, but still there. Sigh, and still no worms feasting on the wonderful repast.

What's going on here? Why aren't they all diving in...wiggling right up and taking big bites? "But wait," you say, "take a bite with what? *No teeth!*"

That's right, worms have no teeth. Like a bird, the worms' crop and gizzard serve as the primary food grinding mechanism. At least birds have beaks to break off pieces!

The worms are not alone. They have helpers in the bin in the form of molds, fungi, bacteria, sow bugs, springtails, grubs and many other micro and macrobiotic organisms. These creatures are the pre-digesters of the banana peel.

Without these organisms, worms would go hungry. These organisms provide the vital service of pre-digesting the organic waste which becomes the worms' food, breaking down large bits of matter into smaller materials.

Respiration

Worms have no specialized respiratory organs. They breathe through their moist skin. Oxygen and carbon dioxide are diffused through the skin to and from the circulating blood stream. Lack of moisture in the worm's environment restricts the breathing process.

---

[6]*Ibid*, page 174.

Prolonged dryness will cause death by suffocation. As we learned earlier, exposure to direct sunlight can lead to death in less than three minutes.

Worm Reproduction and Development

Reproduction

Reproduction dictates survival of a species. Organisms like worms which provide food for many others, must reproduce ample offspring to offset predation. Red wiggler reproductive methods, while simple, are very effective. Most worms are hermaphrodites, meaning each one is both male and female.

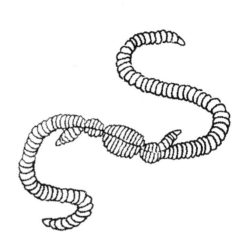

While worms possess both male and female sexual organs, a red wiggler cannot produce offspring alone. Red wigglers must join for successful mating and reproduction to occur.

Some species of worms are parthenogenic, meaning that they can produce offspring without cross-fertilization. *Eisenia fetida* are hermaphroditic but not parthenogenic.

When a red wiggler is sexually immature its body segments are uniform throughout its entire length. As it matures it develops a bulbous gland about one-third of the way down its body

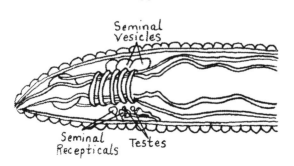

Seminal
Vesicles

Seminal
Recepticals

Testes

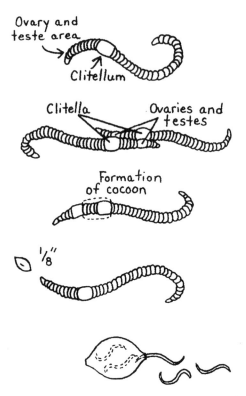

Ovary and teste area

Clitellum

Clitella   Ovaries and testes

Formation of cocoon

$\frac{1}{8}''$

called the *clitellum*. It looks like a swollen band around the body, thus the popular term, bandling worms. The clitellum produces mucus needed for cocoon production.

Two worms come together at the clitellum and use their hair-like setae to hold fast to each other. While joined they exchange seminal fluid. At the same time, grooves on the underside of the worm help transport the seminal fluid to the seminal vesicles for later use. Each worm begins to secrete a mucus ring around itself.

After being joined for up to three hours, the worms separate. The mucus on each worm begins to harden as the worm starts to back out of, or to slough off, this ring. During this process seminal fluid, ovum and amniotic fluid are deposited into the mucus ring. As the ring passes over the worm's head it seals, forming the lemon shaped, yellow-colored cocoon. Over a period of weeks, the cocoon darkens to a beautiful ruby red as hatchlings mature.

Here it must be noted that calcium is essential in the worms' diet. Without calcium, worms will not reproduce. In most cases the organic waste you give the worms has enough calcium in it. But to be sure the worms have an adequate supply, there are three simple ways to provide it:

Population Growth Rates for Red Wigglers

Under favorable conditions your *Eisenia fetida* worm population will multiply rapidly. A mature red wiggler (3 months old) can produce two to three cocoons per week. Each cocoon averages three hatchlings.

⇨ Cocoons take up to 11 weeks to mature and hatch.

⇨ Hatchlings require 2-3 months more before they grow to be mature breeding worms.

Population productivity over 11 week incubation period:
1 worm x 3 cocoons/wk x 3 hatchlings/cocoon = 9 hatchlings/wk
11 weeks x 9 hatchlings/wk = 99 hatchlings/worm

In 2 to 3 months those hatchlings will be mature breeders, producing offspring of their own. At the same time their parents are continuing to mate and create offspring.

As conditions in the bin become crowded, the adult worms will try to leave the bin rather than compete with the young for food.

Population increase is a good reason to harvest the bin every 4 to 6 months.

1) pulverize your eggshells as finely as possible and add them to your bin.

2) grind up a calcium based antiacid tablet and add a teaspoon every week or so.

3) pick up agricultural lime (calcium carbonate) or dolomite (calcium-magnesium carbonate) at your local garden center and add a teaspoon weekly.

Caution: Do **not** use a construction lime such as slaked lime or quick lime as they will poison the worms.

Cocoons

A cocoon is the small, yellow, lemon-shaped object that is produced by red wiggler mating. It is about the size of the letter "O". According to C. A. Edwards and P. J. Bohlen in *Biology and*

46557764777766

Ecology of Earthworms: Eisenia fetida can produce as many as 198 cocoons per year.[7]

Gestation occurs over a period of eleven weeks. The cocoon is hardened mucus and, like a bird egg, contains all the necessary nutrients for the development of the hatchlings. Each cocoon contains between two and twenty potential hatchlings. On average three will emerge.

Climate conditions directly affect cocoon production, for example, when there is too little moisture, the worms will cease reproduction. Peak production occurs when ambient ground and air temperatures are between about 65-80°F or 18-27°C and the environmental moisture content within the bin is between 80-90%. That means moist to the touch but not dripping (like a wrung-out sponge).

When environmental conditions are not appropriate for survival, the cocoons will lie dormant awaiting more favorable conditions. Cocoons have been known to survive for up to three years under extremely dry conditions without being adversely affected.

Hatchlings

It takes five to eleven weeks for the cocoon to mature and hatch. The newly hatched worms first appear as tiny white, thread-like creatures. In approximately eight hours they gain their hemoglobin and change color from white to pale pink to brick red.

Depending on bin conditions, temperature and moisture, hatchlings can take from 53 to 75 days to become sexually mature, that's about two to two and one-half months. The complete generational cycle from one adult worm to the next is anywhere from three to five months.

---

[7]Page 48.

Juveniles

Adolescence is a short lived phase for red wigglers. The length of the worm increases almost daily as their principal function during this phase is to eat, eat, eat! Like many creatures, sexual maturity in *Eisenia fetida* is delayed by cold weather, but not growth rate. Lobsters raised in cold water compared to warm water are a good example. Those raised in cold water grow fairly large before they become sexually mature, while warm water lobsters are smaller upon reaching maturity.

Juvenile worms are distinguishable from adult worms because they look the same from head to tail and do not have the band indicating the clitellum. They are the same color as adults and may be just as large.

Red wigglers spend an average of 56 to 72 days or about 8 to 11 weeks in this developmental phase.

Adults

Adult red wigglers are characterized by the formation of their bulbous clitellum. The presence of the clitellum means they are sexually mature.

Diet plays a key role in the worm's size, which may vary from four to seven inches.. Smaller size does not mean that something is wrong, but smaller worms produce smaller cocoons and fewer offspring.

There is no conclusive evidence that gives a definitive life expectancy for worms. However, a well cared for herd in your bin may give you sterling service for five years or more.

Keeping Worms

Worms are living, breathing creatures. Caring for a pound or two of red wigglers is the same responsibility you take on with a family pet. They will need water, food, and the correct growing conditions in order to carry out their task of reducing the organic waste stream. You can keep red wigglers either in or out of worm bins. We recommend and focus on keeping worms in bins because you can manage the variables more easily. In contained spaces you can be certain that they have enough (but not too much) to eat, are not too hot or cold, are not subject to predators, can be kept at a comfortable level of moisture, and can be harvested easily. The vermicompost is then available to be used however and whenever you wish. You control the amount of worms processing your waste and can share a portion of your herd when they reproduce.

If *containing* your worms doesn't suit your purpose there are other options.

1) Pit-run — Mark off the area to be used, say a 2' x 6' rectangle. Loosen the soil with a pitchfork and remove the first two inches of soil. You can pre-mix organic materials such as manure, peat, household food, or soaked shredded paper into the soil and water well, then add your worms. Or you can just water the area and add worms.

Always water the area well before introducing the worms. Add them in the late afternoon while there is still enough sunlight to encourage them to dive for cover, but not enough strong light to damage them.

Feed the worms as you would in a bin. Cover them with straw or soaked sheets of newspaper to help retain moisture and minimize predators. You still may have visitors of an animal nature interested in your scraps or in a meal of delicious worms.

2) Trenching — Mark off the area and loosen the soil as instructed with the pit-run. Remove the soil to a one foot depth and add back food and bedding moistened and mixed with soil.

3) Windrow — is a technique used by professional worm growers where bedding materials are piled on the surface of the ground in long rows, e.g., 2' x 10', 4' x 20' or larger.

Mark off the area, loosen the soil, water well and lay out a layer of bedding material at least six inches high. The six inch depth ensures enough depth for worms to escape if the birds discover your cache.

Bedding material can include pre-heated compost; manures from plant eating animals such as rabbits, horses, llamas, chickens, etc.; mixed with soaked shredded paper or straw as examples.

Add your worms in the late afternoon and feed as usual.

Because the windrow has so much exposed surface area, water loss is a problem. If you use this method, try laying soaker hoses down the length of the windrow for easy water care.

As with the pit-run method, you can cover the area with soaked newspaper or straw to help retain moisture and deter predators. Do *not* use plastic, you will overheat your herd!

4) Tree base[8] — still experimental, but the University of Oregon is testing vermicomposting around fruit trees. The area is cleaned and raked out to the drip line of the tree, food wastes are applied in a thin layer every 14 days and well watered. The area is sprinkled with rock dust then covered with leaves. There were no worms evident at the beginning of the experiment and none were added. Now trees and vegetable flourish with many worms evident in the soil. The worms will stay in the area as long as there are ample organic food sources and moisture available.

I want to keep worms, how do I get the right kind?

Most cities don't list worms in the yellow pages and most of the worms you dig up in your garden are not the efficient composting worms. Try calling the County Agriculture Department, University Extension, Master Gardener office, or local garden center for advice for your area. We have included some additional sources of information in the appendix.

---

[8]*Worm Digest*, Issue 11, pages 19-21.

Setting Up a Worm Bin

Conditions that control worm populations are the same as those required for them to survive in a contained space:
1) ample, but not excessive food
2) sufficient surface area
3) population density
4) nutrition for reproduction
5) moisture
6) ambient air and ground temperature, 55 - 85°F (13-27°C)

What kind of container should I have?

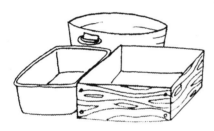 The type of container you use for your worm bin is limited only by your imagination. Below are a few guidelines you'll want to keep in mind. Also, remember, a lid will keep it dark, help retain moisture, and keep the birds from feasting on your herd. Even a bucket can be used as a worm bin.[9] When selecting a bin, it is important to pay attention to the square foot surface area listed in the chart on worm to waste ratios.

The consider the following in selecting your worm bin:
❑ The inside of the bin will always be wet, so cardboard and similar materials are not appropriate.
❑ Select a size that is easy for you to handle. If you maintain a bin kept in the house, it will need to be taken outside before harvesting.

---

[9]Mary Appelhof, *Worms Eat My Garbage*, Flowerfield Press, page 11.

❑ Appearance is important to most people, so obtain or build something that is attractive.

❑ Plastic holds up well and retards moisture loss. Plastic containers often come with the necessary lid. Caution: worms prefer to work in the dark, so try to avoid clear plastic or keep the container covered in the shade.

❑ Wood breathes well but warps and rots. It is also much heavier than plastic, especially when filled with worm castings.

❑ Wood manufacturers infuse some wood with pest resistant chemicals to increase its longevity. These chemicals can be poisonous to the worms.

To reduce cost and keep with the spirit of recycling, try to use a container that has had a former function. You may build one of scrap, have one at home, or discover the perfect worm home for pennies on the dollar at a second hand store or garage sale.

Some suggestions:

✓ old dish pans are a good size for an individual or couple.

✓ plastic flip-top file storage bins work wonderfully.

✓ bins made from *safe* (not pressure-treated) scrap wood are effective.

✓ recycled barrels, cut in half lengthwise and bolted together at the mouth provide inexpensive bins with a lot of surface area.

Clean the container thoroughly before you add worms. If your bin doesn't have a lid, a piece of wood or cardboard with a rock on top will do.

Remember: No matter the type of bin you choose, recycling is part of the goal!

What is the right size container for my worms?

Take a week or so, look at or weigh your waste recyclables from the kitchen. Compare the volume of your waste to the chart below to help determine the right size bin for your needs. As a rule of thumb, for a one to two-person household a twelve to sixteen quart bin will work. To meet the needs of a two to four member family, a twenty-two quart, flip-top bin will do the job.

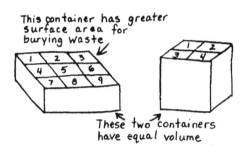

This container has greater surface area for burying waste

These two containers have equal volume

The container doesn't have to be very deep, remember, E. fetida are top feeders, anything deeper than a foot is wasted. The more surface area the better.

Measure the length and width of the container in inches. Multiply the length by the width to calculate the surface area of the container in square inches. One square foot equals 144 square inches.

The ideal beginning volume of worms for different size bins is given in the chart. Once the bin has been set up the worms will begin to multiply. In about four months the volume of worms in the bin will have tripled. When the volume reaches three-quarters pound per square foot, the worms become overcrowded and will try to begin to migrate. Worms on the sides and top of the bin are a good indication that it is time to harvest the castings and divide the herd.

How much and how often do I feed my worms?

In ideal conditions in the wild, a worm will eat its own weight daily. In the confined space of the worm bin you can

Ideal Surface Area to Worm Volume for Starting a Worm Bin	
Surface Area (sq.ft.)	Worm Volume (lb.)
1	¼
2	½
3	¾
4	1
5	1¼
6	1½
7	1¾
8	2

expect a worm to eat up to one-half its body weight daily. Worms burn up less energy in a worm bin. They do not have to range very far to locate food and they cannot travel far in the confined space.

A family of four with a twenty-two quart bin, starting with two pounds of worms, can expect them to consume one pound of waste daily. Estimate processing three to seven pounds of food per week. Feeding the worm herd twice a week should be fine.

How do I figure out the volume of my organic garbage?

An easy way for you to find out how much garbage you generate in a week is to enlist the help of family members to measure it.

Children love to get involved with worms. Create an enjoyable ecology project for your children, grandchildren, or borrowed children from your neighborhood.

You and this willing group will need to obtain several collection containers and a scale.

In one container collect kitchen scraps such as coffee grounds, tea bags, orange rinds, pancakes, chicken fat, yogurt, etc.

In another container collect paper waste such as paper towels and toilet paper rolls, napkins, paper plates, tissue, and any other non-food organics. You can put these in the worm bin too,

Required Worms for Waste Volume

Pounds waste/day	Pounds worms
½	1
1	2
1½	3
2	4
2½	5
3	6
3½	7
4	8

just soak them for at least 24 hours before placing them in the bin. Soaking helps aid in the decomposition of paper improving its quality as worm food. To prevent the food collection from rotting and producing unpleasant odors, weigh your waste daily. Record the amounts and store this waste in the freezer until you are ready to feed the worms. Freezing allows the organic material to take up extra moisture and begins the molecular breakdown of the waste. If you don't have freezer space, store your waste in an airtight container under the sink or outdoors in a shaded area until feeding time.

Using the scale weigh your waste for one week. We are assuming this is an average week for your household with no extra guests or dinner parties!

Let's say your household produces one pound of organic waste every day, or seven pounds per week. We know worms consume about half their weight in food per day, so you will need two pounds of worms to process this volume of waste.

How can I be sure that the worms get the air and oxygen they need?

Drilling holes in the bottom of the container provides air circulation and necessary water drainage. The number of holes in your worm bin depends on the size of the bin you choose. Place the holes at least two inches apart using a ¼" drill bit. For example, about fifteen holes will go in the bottom of a 12" x 14" dish pan. Put the bin on two by fours for drainage and over a tray to catch the drips. By keeping a pan under the worm bin, you can capture the runoff, a nutrient rich liquid called *worm tea*, for your plants.

RECIPE FOR A WORM BIN

EQUIPMENT AND INGREDIENTS - WHAT YOU WILL NEED
- [ ] one container for worms, with lid
- [ ] one old sheet, cloth, or other porous liner
- [ ] one electric drill with ¼ inch drill bit
- [ ] bedding: shredded newspaper, corrugated cardboard, coconut fiber (soak at least 24 hours before assembly)
- [ ] tray to catch drips beneath worm bin
- [ ] two short sections of 2" x 4" to raise bin off the ground
- [ ] generous handful of garden earth
- [ ] location in deep shade for bin
- [ ] composting worms – red wigglers – as many as you need to compost your waste volume

INSTRUCTIONS

Fold the newspaper sheets in half and tear into long strips about ½ inch wide. An average daily newspaper should weigh about nine to fourteen pounds when wet, sufficient for up to a twenty-two quart bin. Shiny newsprint is fine as bedding.

Soak the newspaper strips overnight. . .at least twenty-four hours. Chlorine and other chemicals are put in our water supply for health and safety reasons. Soaking overnight allows the paper to become fully saturated and allows these chemicals to evaporate so they do not harm the worms.

Invert your bin, place the liner material (cotton or nylon) over the container and mark four inches down all sides. Cut the liner. The liner will fit into the bottom and extend up the insides of the bin preventing worms from escaping. Nylon won't breakdown as quickly as the cotton and will last longer. No matter what material you use, try to have the lining extend up about four inches on all sides of the bin. Wet the liner, wring it out and place it in the upright bin.

Wring out the long pieces of soaked newspaper. It should be about as moist as a wrung out sponge. Start filling the bottom of the bin, fluffing up the newspaper as you go. Fluffing the paper allows the worms to crawl through it easily, remember these worms are not burrowers!

Cover the bottom completely with the newspaper bedding at least four to six inches, deep enough to secure the liner. Think of this as the worms' playground, if the bedding isn't fluffed the worms won't be able to slither around.

Next, sprinkle in a handful of regular garden dirt; real dirt, not soilless planter mix. The worms will use the sandy soil particles to grind up the food in their crop and gizzard. If they can't grind it up they won't get their proper nutrition. As mentioned before, the bacterial helpers in the soil are essential to worms for pre-digesting organic waste. Keep in mind that the more diverse the organic wastes you feed your worms, the richer their castings will be!

Put in your worms and close the lid to give them darkness and shelter. Let them rest and become accustomed to their new home for two to three days, then start feeding them. Many people feed their worms daily, however, two to three times a week is fine.

Use a garden trowel to push back the bedding when you add food waste to the bin, then cover it over with the bedding burying the new food waste. This will reduce odors and make the bin less attractive to roving animals. Devise a chart of your own or use the one illustrated to keep track of your burial waste sites allowing the worms to process the food before you go back to introduce more. An easy way to do this is to mark the lid of your bin with a waterproof pen for the days of the week and not go back to that site until all the waste is gone.

Excited?! We bet you are, because now you have a home for your new worm family!

Shredding the bedding is time consuming, isn't there a better way?

Shredding paper for bedding doesn't have to be grueling manual labor. Personal paper shredders are becoming less expensive and quickly shred up ample quantities of newspaper. Or, if your are resourceful, free shredded paper is all around! In this day of automation and secret files, school districts, banks, offices and hospitals all shred their paper. Call and ask if you can pick up a bag.

Some people use peat moss in their bins as a worm bedding instead of newspaper. This is a good fiber material for worms, but check your sources and content carefully. Acidity levels of peat can vary greatly according to plant content and source area, some will be too acidic for worms.

While some environmentalists decry the harvesting of peat as a non-renewable resource, particularly in Europe, this also varies by source. Look at the label and find out what levels of wetlands protections are provided by the country of origin. Canada, for example, has vast resources and never allows use of more than .02% of available area to be utilized. Like plugs in a lawn the area is allowed to regrow before further harvest.

Another bedding material is shredded coconut fiber, a renewable byproduct from the food industry. It comes in blocks that expand, absorbing up to ten times its weight in water. It should be available in your local gardens shops or check the resources list at the back of this book for mail order catalogs which carry the peat-free bricks.

Depending on the place of origin and the processor, coconut fiber may be high in salts. The industry is still too new to have consistent quality control. Check each new batch you buy for salt content. If it is high in salts, soak the material for at least 24 hours, discarding the water, and run water through it afterward to leach out the worm-deadly salts.

How often should I add bedding to the bin?

You should only need to put the bedding in the bin when you begin and each time you harvest your worms. If the bin is too wet, however, you may add shredded paper or cardboard to soak up some of the moisture. It takes the worms about three to four months to eat their supply of bedding and you will want to harvest them by that time.

Should I add water to the worm bin?
Much of what you put into the worm bin will already have a high water content. The contents of the bin need only be moist to the touch, (remember the wrung out sponge) so the worms can slide by each other.

If you freeze scraps, when they thaw in the bin they will supply the worms with all the moisture they require. If you live in a very hot, dry climate you may experience excessive evaporation. If the bin contents feel dry to the touch and you think the worms need more water, put some tap water in a container overnight to release the chlorine. Add water to your worm bin sparingly. By keeping the bin covered you have created a mini-ecosystem. The bin's own condensation should be the only moisture needed.

Congratulations! You no longer need your electric garbage disposal!

Feeding Earthworms – Do's and Don'ts

You've set up your bin, the worms are in, for the next few months all you have to do is to feed them the right things in the proper amounts until it is time to harvest the castings.

What do they like? Organic foods, just like all of the things people eat, only with even greater variety. *Organic* includes everything that is alive or ever was alive. Just like humans, *Eisenia fetida* appreciate plenty of fiber and a varied diet. They don't have to be told to eat their vegetables, in fact, those are some of their favorite foods.

Keep in mind that you are not only feeding the worms, but also thousands of microscopic organisms which are breaking down the food for the worms. These organisms can only work on the food surfaces that they can reach, so the finer you chop up your kitchen waste, the easier it is for all of the decomposers to do their jobs. But if you don't, eventually they will break it down but it will take longer.

Killing worms with kindness

One of your most difficult tasks will be to restrain yourself from overfeeding your earthworms. Temptation abounds and it is hard to resist.

It's summer and the family got together with the neighbors for a picnic and watermelon fest. What do you do with all those rinds, the leftover potato salad, hot dogs and hamburgers the kids didn't finish? Or, you just finished peeling and pitting a full lug of peaches to can them for winter desserts, here you are with a huge bowl of trimmings. Now, it's Thanksgiving and it was your turn to cook, wouldn't the worms just love these extra treats?

Well, maybe they would, but not all at once. You have several alternatives:
1) freeze the excess to gradually it to feed the worms later.
2) place it deep in your active compost pile.
3) dig a hole and dump it in, worms in the soil will find it.

4) if you live in an apartment and haven't any other choice, put it in the trash or turn on the disposal.

Probably more home worm herds are lost to overfeeding than almost any other factor. Worms in the wild don't have this problem because they are free to roam. If there is a big body decomposing in the woods, they can wait until it is their turn in the food chain and come and go at will. In the worm bin, if excess food is anaerobically decomposing, there is nowhere to escape.

If you find you consistently have more food than your bin can handle, get a larger bin or start a second one. There isn't any rule that says you are limited to only one.

What to Feed Worms

You shouldn't run out of things to feed worms, many of the same organic materials you put in compost bins can be fed to them. Try different items to see if your worms like them. The worms will let you know what they prefer.

Many guides will list animal products as not being good worm food. The worms will definitely eat them, but they are also attractive to larger animals and can cause odors as they decompose. This takes a little extra care. Remember: with dairy and meat products, burying will help keep the critters from becoming a problem and keep odor to a minimum.

Since you know we no longer use our garbage disposal you know we put in oils, fats, chicken carcasses, cheeses, cereal and everything else. We haven't had any problems with our worm bins and these usually designated as *no-no* wastes. It is up to you to make those decisions.

If you are unsure about something start with just a little of it. Don't be afraid to try combinations of different items.

Here are a few examples:

apples and peels   cake
baked beans   molasses
vegetable trimmings   pears
biscuits   celery
cabbage   leaves
oatmeal   corn cobs and husks

coffee grounds and filters
cheese
cream of wheat
cucumber
table scraps
deviled eggs
egg shells
farina
tomatoes
turnip greens
melon rinds

grapefruit and peels
grits
lemons
lettuce
onion peel
pancakes
pizza and crust
bread
potato salad
cream
leftovers

The list is endless. Remember to bury the food in the bedding and to not overfeed the worms.

Beware - Foods Potentially Dangerous to Worms
A little of anything is probably okay but too much may be toxic. We mentioned before that diversity is an important element in the worms' diet. This list will acquaint you with some of the potential organic foods that may generate problems if given to the worms in quantity. Again, the list is not comprehensive but tells you the reasoning behind what foods to look out for and avoid putting in your worm bin.

❏ Wood products treated with chemical preservatives or "pressure-treated" — the active ingredient is cyanide, even minimal quantities are deadly to earthworms. Wood takes a long time to break down unless it is in very small fragments.

❏ Plant material treated with insecticides or weed killers — again, know your source and avoid all of these materials, even small quantities may harm your worm herd. Old references indicate pesticides to eradicate "pests" in the bin. We do not suggest it.

❏ Carbon paper — avoid this, toxic inks are used to make it.

❏ Salty snack foods like chips, fries, olives — avoid foods with salt granules or high levels of salt. Soak salty foods in water for at least 24 hours before adding to the worm bin. Dispose of the salty water elsewhere.

❏ Manures — feed lot animals are fed huge quantities of salt to bulk them up before slaughter. Make sure all herbivore animal manures are composted before adding to the worm bin or they may be too hot (contain too much ammonia) or have too many salts for good worm health. A good rule of thumb is to allow manures to sit out in the elements for at least a year to allow rain to leach out excess urea or salts.

❏ Pet (cat and dog) feces — Dispose of this waste and do not feed it to the worms or compost it. May contain viral or bacterial toxins, also attracts vermin. The same is true with all other carnivore (meat eating) feces.

❏ Citrus waste — a few orange peels are okay, but if you squeeze fresh orange juice every day it is best to compost that waste rather than applying it to the worm bin. The rinds are very acidic and create an imbalance in the pH of the worm bin if applied in quantity.

❏ Vinegar and vinegar based salad dressings — again the vinegar changes the pH of the bin creating an acidic condition. A little dressing on uneaten salad greens is okay.

❏ Meats and other animal byproducts — these will give off a foul odor as they decompose and will attract vermin. Avoid adding too much animal protein to a worm bin and bury it in the bedding to reduce odor.

❏ Green grass — especially in large quantities, green grass decomposes thermophilically creating very high temperatures harmful to worms. It is better to compost the grass with other foods and feed it to the worms after composting.

❏ Alcohol — forget it, very bad for worms.

❏ Fertilizers and other chemical compounds — most fertilizers are chemical salts which, while they provide nutrients to plants when they break down in the soil, are much too strong for worms.

☐ Diseased waste from plants or animals — vermicompost does not heat to high temperatures to kill pathogenic organisms, likewise seeds are not rendered infertile.   Avoid these materials if you do not want them to spread.

☐ Fruit pits — these aren't bad for the worms but you may be disappointed to see them solid and whole in your bin after all else is composted. Nature gives seeds protective qualities which allow them to stay in the soil until it is time for them to germinate. If seeds were good worm food we wouldn't have new plants to feed them (and us) in the future.

Use your own good common sense. Worms need a fiber source like newspaper or other bedding materials, so always use those in setting up the bin. However, when feeding your worm herd, if you are in doubt and wouldn't eat it yourself, why give it to the worms?

Things Happen

While we would be thrilled if you read our book, followed the instructions and took perfect care of your worms, you will probably run into a few snags. Don't give up! Just like any new endeavor, we all learn from our mistakes. Never let lack of experience keep you from trying something new.

We hope the following true stories will help you overcome any reservations about trying worm composting.

Population Boom

We all want plenty of worms in the bin. At times you will find surprising increases of population. Sometimes we conduct experiments to see if there are better ways for worms to process our waste.

When we went out to feed the worms one morning, *all* the adults were massed at the top rim and on the lid of the twenty-two gallon flip-top bin. They looked fine, active, the bin smelled okay but the worms are trying to tell us something. Satisfied worms don't try to escape. Something was wrong in the bin. We became detectives to find why the worms' behavior changed.

- Did we over-feed the worms?
- Was harvest time approaching and food supply running low?
- Was a sprinkler left on filling the bin close to flood stage?
- Was there too much citrus waste making the bin too acidic?
- Did we add something with high salt content?
- Did many cocoons hatch, over-crowding the bin for the adults?
- What had we done to change their environment?

In our case, we had been testing the use of fresh grass as worm food in the bin. In the process, we did several things wrong:
1) we added too much grass.
2) we used donated grass and didn't check the source
3) we didn't find out if pesticides or weed killers were used on the grass.

The grass began thermophilic (bacteria generated, high temperature) composting in the bin, overheating the worms.

Experiments are positive. We gained valuable knowledge from this one. Here are a few of the things the *grass-in-the-bin*, showed us:

✓ Always know where your organics come from. Now we ask *donors* who volunteer wastes for the bin what has been sprayed, poured or coated on the waste?

✓ Grass is used either as green manure or regular compost material, not put with worms.

✓ Pay attention to the bin conditions: Wet? Dry? Has an unpleasant odor?

✓ When experimenting, have a worm annex (another bin) ready to go, with bedding and a liner, just in case.

While we no longer add fresh grass to worm bins, we do add grass seed, inexpensive bird seed or seeds from the kitchen sprinkled on top of the material in the bin. The seedling roots help to aerate the bin contents. Aeration reduces odor and helps neutralization of the pH level. The germinated seeds are turned under and die, becoming a food source. We make sure there are never too many seeds sprouting and dying to create a condition for spontaneous composting.

All was not lost trying to put grass in the bin. But we learned to keep trying different wastes, and keep a close eye on the herd when we try something new. Be adventurous, innovative and creative. Above all, be kind to your worms!

A Change in Scale:
Problems

Our first attempt at managing a large worm herd nearly ended in disaster before it got underway. We had already kept worms for many years at home and thought we knew a lot. We were implementing a program at three elementary schools in our community to teach recycling and sustainable agriculture through vermicomposting and regular composting.

Our folly began with the miscalculation of the food waste being generated at each school. We "guesstimated" the weight of food collected in a week, rather than weighing it. Garbage is deceiving! The result was a grossly underestimated total.

Using our "generous, guesstimated figure" we calculated the amount of worms needed to handle our waste based on the knowledge that worms will consume half their weight daily in captivity. All seemed to point to a green light.

With worms in place and waste separated from non-food items, we began feeding. After the third day steam began rising from the beds. This was not a brisk, frosty morning but a hot, dry day in southern California. Worms huddled in writhing masses in the corners trying to escape the heat.

Analysis

What was going on? Why weren't the worms eating? After careful analysis, we determined that three things were happening:

☞ Volume — Our total pounds of waste was two to three times greater than the amount we had calculated. Don't eyeball your garbage when determining how much your household generates, actually weigh it. Because of the gross difference in the pounds of waste generated, we had overfed the worms.

The high volume of waste was thermophilically composting in the worm beds increasing the temperature in the bins to well over 100°F (38°C). In other words, bacterial action was heating the waste before it could be broken down into simple enough forms to provide food for worm consumption. The high temperatures generated were too extreme for the worms to survive. Poached worms aren't a pretty sight!

Not only were the temperatures excessive but ammonia gas being released from the bacterial action can also be toxic. In compost bins it is a signal to add carbon materials, such as the newspaper used for bedding.

☞ Type of food — A second problem due to overfeeding was excessive proteins in the foods being fed to the worms resulting in protein poisoning. The high levels of meats and proteins were far above their normal diet. Their bodies were literally exploding from the excess protein they consumed. It was a dramatic example of dietary imbalance.

☞ Chemicals — The third unexpected difficulty resulted from the high levels of nitrates, sulfides and other preservatives used in the school food. These chemicals serve their purpose, but in high levels were too strong even for worm gastrointestinal systems. Just like when you eat too much salt it makes you sick. Preservatives were a contributing factor to killing our herd.

Corrective Actions

We stopped feeding the herd immediately. We mixed fresh bedding into the bins to stop the food from composting. At the same time, we iced the worms down daily for a week using fifty pounds of ice in each bin, twice daily; also adding fresh, dry bedding to soak up the excess water from the ice. By the end of a week we managed to cool down the bins and save part of the herd.

Once the composting action was halted, the worms were kept under careful observation until we could begin feeding again. Correcting excessive feeding eliminated the composting problem. Monitoring the levels of foods introduced into the bins corrected the protein poisoning as well. If you think about it, the major portion of what worms find naturally in soils are remnants of plants, leaves and the like. It is exceptional when a worm encounters a protein source from a higher animal and never in the quantities we put in the bin.

But what do you do with the excess food? We process it separately, composting it with grass clippings from the campuses, creating a fabulous food supplement for the worms during vacations and holidays. As a side benefit grass in the compost helps to control odors, as will any brown leaves and shredded paper.

Battling preservatives was a dilly of a problem! When the worms continued to die we were really at a loss, but only for a short time. Students and parents had long complained about the quality of the pre-prepared foods in the school lunch program. The meals left a lot to be desired.

So, into the dumpsters we went, on a hunch, to do a bit of scrounging around, also known as field research. And what did we find? Labels with ingredients that would mean eternal preservation for anyone consuming those products, in any amounts, over periods of time, among them nitrates, nitrites, sulfates, sulfites and more.

We also discovered a way to make enemies lickety-split. We presented our findings as to the ingredients and results of soil samples from the worm bins to school authorities. Consider: if the preservative level was so concentrated that worms were dying from it, what are the effects on our children? We were not in line for Dale Carnegie Awards.

Conclusions

The end of this story is also threefold:

✓ We weigh **all** the waste before feeding it to the worms.

✓ After corrections and adjustments (including reducing the meat levels) were made to suit worm nutritional needs, the worms have thrived.

✓ The issue of frequency and levels of the preservatives in the food is still on the table, but food quality has improved. It will take long term incremental betterment. We care deeply about all these things but must deal with some one step at a time.

Mighty Mites

Our editor ran into this problem and we were thrilled when she solved it herself. One indicator of a too-wet bin is the presence of tiny red mites. Each one seems about the size of a period or a little larger. They will mass over the surface of a bin in large numbers covering the food waste. At one time people thought that mites would harm the worms, but it just turns out that they are at the feed trough like all the rest of the creatures. However, they are not very attractive.

She wracked her brain because she never added any water to the bin, relying only on the moisture of the food, yet it always seemed too wet. Finally, she was watering her plants and it struck her like a gong. As she watered the potted tree next to the bin the water splashed over the bin. Because she had used old instructions to add holes in the top of the bin for aeration (no longer recommended), the splash-over from the tree was moistening her bin every time she watered.

Her first action was to open the bin during the day to dry it out, placing the lid bottom-side up as well. The mites on the inside of the lid and on the surface were soon killed by the

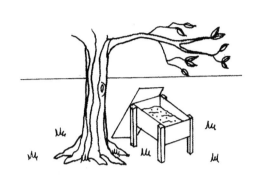

sunlight. She just brushed them off the lid. The bedding was deep enough that the worms dove for cover and weren't affected adversely by the drying out period.

It was almost time to harvest the bin anyway, so she did and set the worms in their drier quarters. Rather than using the worm compost immediately, she decided to add it to an already active compost bin to allow other creatures to feed on the dead mites. another alternative would have been to bury it in her garden. She moved the refreshed bin to a shady spot away from where she waters.

If you have had mites and want to find out if you have fixed the problem, place a piece of white bread on the surface of the bin food overnight. In the morning check to see if there are any red mites on it. If there are, you either have too much food in the bin or it is still too damp.

Ants are Not Relatives

Ants are part of nature. Their presence in the worm bin is an indication it may be too dry. They love a free food source and especially enjoy unburied food. While you could just leave them alone, if you don't like their presence there are several ways to discourage this unwanted horde without resorting to toxic chemicals. When you draw a line in the sand, if it is the right kind of line, they will seldom cross over.

Put the worm bin over a large tray, elevated on bricks or two-by-fours. Fill the tray with water creating a raised island. The worms will reside peacefully in their moated castle.

Or, put a bead of petroleum jelly completely surrounding the the bottom and top of the bin about two inches from the edge. A line of baking soda surrounding the bin will also discourage ants (water will flush away this line of defense).

Don't use pesticides around the bin or you will put your worms at risk.

Air, Food, Water - Necessary to Life

Sometimes the bin will begin to take on a sour odor, worms will be sluggish and seem to be dying. One or more conditions exacerbate this situation: not enough free air circulating in the bin, bin contents are too wet, or there is more food than the worms can handle.

To increase air circulation, first check to see that the holes in the bottom of the bin have not become clogged. Aerate the bin by leaving the top off during the day for a few days. (Don't forget to close it up at night if there are predatory animals in your area.) Opening the bin will also help to alleviate an overly moist condition, provided it isn't raining.

To reduce moisture level, you can add more dry shredded paper for bedding. Stir the paper so that it is well mixed into the bin to soak up the excess moisture.

For overfeeding, stop feeding the worms for four to five days. Check the bin to be sure the old food is being consumed. Freeze the waste to feed the worms later, or divide the worms and start another bin.

Fruit Flies

Fruit flies can be a problem, especially in the warmer months. If they bother you, try keeping the lid on as much as possible. These pests are a good reason not to have holes in the lid or sides of the worm bin. Bury the food waste in the bedding so they won't have easy access.

Pour a cup of apple cider vinegar in an open container and set it beside the bin. It will attract the fruit flies and they will drown. It is important to not put the container in the bin, if it tips over the acidity will kill worms.

Another effective fruit fly trap can be made by recycling the plastic lid from a coffee can, cottage cheese container, or other plastic container. Put a hole in it and use a piece of coat hanger to hang it nearby. Smear it all over with Tanglefoot™ or petroleum jelly making a no-cost free fly trap!

Molds, Fungi, Yeasts

This is only a problem if someone who helps to care for the bin has allergies. Very few people are affected by the natural breakdown of waste food in the bin. The allergy affected person can be in charge of other activities such as harvesting, weighing the waste and castings, or using the castings in the garden.

None of these minor setbacks will bother your worms or you, if you take your role of worm manager seriously and pay attention to what is happening in the bin. Many other micro and macroorganisms share the bin with the worms. They all have jobs and are necessary for a healthy bin. The only difference is that you are in charge of the environment and must keep them in harmony with nature's cycle of life.

As you see, even those of us who have experience with worms make mistakes. It is part of the learning process. The errors taught us that all was not lost and to use our brains to correct a problem.

We feel sure you will do a good job managing your worm herd.

WORM BIN TROUBLE SHOOTING GUIDE

Symptoms	Problems	Solutions
Foul smell	Lack of air	Fluff bin contents
		Sprout seeds in bin
	Composting food	Cut back on feeding
		Bury food waste
	Too much water	Open bin to dry out
		Add dry bedding
		Check drainage
Dead or whitish stringy worms	Starving worms	Feed worms
	Overpopulation	Harvest bin
	Too wet or dry	Add dry bedding or water
Fruit flies	Exposed food	Bury food in bedding
	Occur naturally	Apple cider vinegar trap
		Fly trap

Ants	Easy access to food Occur naturally	Bury food Run bead of petroleum jelly around bin Set bin over water
Mites	Bin too wet	Open bin to dry out
House flies	Exposed food	Bury food Make fly trap
White thread-like worms (*Enchytraeids*)	Occur naturally	Ignore them, part of the life-cycle of the bin
White, yellow, orange matter in the bin	Occur naturally	Molds, yeasts and fungi, part of the bin life cycle
Seedlings growing in bin	Good sign	Will aerate the bin and keep castings loose
Other creatures in the bin	Good sign	A healthy bin mirrors nature.

Special Situations

I'm going on vacation, do I have to hire a worm-sitter to feed my earthworm herd?

We mentioned before it's a great idea to freeze your excess leftovers. Save enough out to cover the amount you would feed the worms while you are gone and bury it at one end of the bin. Use a thin layer of soil or additional bedding to cover the food and you can go away with a clear conscience. With ample food the worms can take care of themselves for two to four weeks. Or, have a friend or neighbor get involved and soon you may be sharing a common interest!

How do I winterize my worm bin?

There are many ways to winterize your bin depending on where you keep it. If your bin is in the house and your home is heated, you have no worries. On the other hand if your bin is exposed to severe temperature changes, in the cellar, garage, tool shed, or under a tree, insulation needs to be provided. Insulating a bin can be easy and inexpensive using materials such as: 4" sheet styrofoam, hay bales, newspaper or building insulation. Use your creative skills and recycle.

The bin must be totally surrounded by the insulating material,

(don't forget the bottom). Use four to six inches of good insulation for moderate winter weather. If your bin is outside in consistently cold and snowy weather (20°F or -7°C and below) you should consider full sized hay bales on all sides. We recommend moving the bin indoors to the cellar, garage or under the sink. Remember, worms won't eat as much if they are cold.

I live in an apartment, can I still keep a worm bin?
By all means! Many families in Vancouver, British Columbia are participating in a city-wide worm composting project and keep worm bins under their sinks.[9]

Another excellent example of worm bins in the home can be found in India. The Shrikhandes' household use their same vermicomposting bins to raise crops to supplement food for the family. Nearly twenty years ago, Lata Shrikandhe said to her husband, "If we had an underground floor, I would have a pit (for worms)." He replied, "Why not on the terrace?"[10] Ever since the vegetables are grown in the same raised beds where the family is vermicomposting their organic food waste.

You can keep your worm bin in any convenient place, as long as it meets the worms' living requirements. Creative vermicomposters are ingenious and been known to use worm bins as furniture: a bedside nightstand, a living room flip-top window seat, as a cabinet underneath an aquarium, or as a front porch or patio seat.

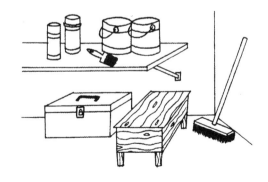

---

[9]See Appendix for further information.

[10]*Worm Digest*, Issue 15, page 16.

Harvesting Worm Castings

How long can I keep feeding the worms in the same bin?
Mark your calendar because you will need to harvest the worm castings in four to six months, and every four to six months after that. The rich worm manure, called castings, is one of the best soil amendments for your garden. It is the worms' thanks to you for your good garbage.

If you wait too long to harvest castings, they will become hard-packed. Natural gases resulting from decomposition will eventually kill the herd.

As a preventive measure, take some inexpensive wild bird seed and sprinkle about one tablespoon over the top of the bedding. Don't add water, there will be plenty of moisture for the seed to sprout. Any type of fast growing seed will do, wild bird seed or seeds from the foods you eat. As the seeds sprout, roots reach down aerating the castings. This allows air to flow freely within the castings.

The worms and plants exist in symbiotic relationship. The green plants utilize the carbon dioxide dispelled as waste by the worms and give off oxygen. The worms breathe in the oxygen given off by the plants. The roots loosen the castings and utilize the nutrients in the worm manure for growth.

Hard packed castings and deadly gas are not the only reasons why you will need to harvest your worm bin. As the cocoons mature and begin to hatch, adult worms migrate away from them to avoid competition for food with the immature worms.

These young worms, like teenagers, have voracious appetites. You want to keep a vital, healthy worm bin to assure continuous worm reproduction to efficiently process waste.

When the worm population burgeons and bins become over-crowded such that conditions in the bin are unfavorable for the worms, they will try to migrate. Signs of worms attempting to migrate are:

1) large numbers of worms on the sides and lid of the bin

2) worms escaping out of the bottom of the bin
3) stressed worms that can't escape massing together
4) in extreme cases worms begin dying.

How do I harvest the castings?

Harvest worms on a sunny day. The day before you are going to harvest your worms, tear up or shred newspaper for new bedding. Soak it twenty four hours, like you did before, so you can immediately put your worms back in their refreshed home.

We must protect the worms from ourselves—the natural oils in our hands can destroy the fragile shell of the worm cocoons— so always wear gloves when working with worms and worm castings.

To prepare to harvest you will need:
☐ an old tarp or sheet,
☐ a container for the worms and
☐ another container for their castings.

Lay the tarp or sheet on the ground or on a large table. Gently invert the bin, making four or five cone-shaped stacks with the contents.

Let the casting cones sit for about ten minutes while you clean the bin and reline it with moist cloth. Add the newspaper, fluffing it up so they can crawl through. Don't forget to add the handful of dirt — grit for the gizzard!

Every ten minutes take off the top two inches of castings and put them in a pail. Just use your gloved hand or any flat-sided scraper or stick. We always have plenty of other garden projects to keep us busy between the ten-minute scrapings.

The worms will dive for cover each time you scrape. When you are down to the bottom two inches, you will have a mass of worms. Your harvested castings should also have some stray worms and hopefully lots of cocoons. That's good, because you will need them if you are going to cure your castings.

Put your starter amount of worms back into the bin to begin your vermicompost project again.

With the rest of the worms you can start a second bin, give some to friends to start their own bins, or release the worms in your garden. If you decide to release the worms, make sure there

Pour bin contents onto plastic sheet.

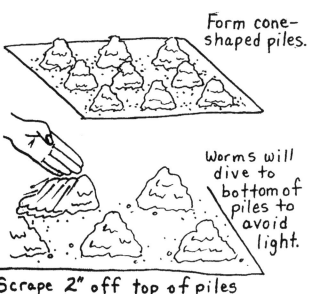

Form cone-shaped piles.

Worms will dive to bottom of piles to avoid light.

Scrape 2" off top of piles every ten minutes.

Eventually, all that's left are the worms.

Place worms and vermi-compost in seperate containers.
Store vermicompost for use in garden, houseplants, etc.

Add worms to box of new bedding.

is plenty of organic material in the area for the worms to eat, such as leaf litter, mulch or partially composted material. Do it in the late afternoon in an area that you have generously watered.

We really like sharing our worms with others. There is nothing like hands-on experience to see how easy and fun it is to have worms. Have your friends bring over their own already-prepared worm bins on harvest day. Now you have the expertise to let them know if they are doing it right. They can see for themselves how they, too, can reduce our waste problem while getting a wonderful garden amendment for free!

Some people like keeping a journal of their results of vermicomposting. Just for fun, you might weigh your castings to see what you and your worms have produced. Weigh the worms, too.

Passive Harvesting Technique

There is a slower method of harvesting your worms and worm castings if you are very patient. To have worms harvest themselves, first soak shredded paper like you did to set up your bin. The next day move all of the finished vermicompost to one side of the bin. Place the damp paper in the empty side of the bin, fluffing it up for the worms.

Bury your food waste in the paper as you have been doing, but only at the far end of the bin. The worms will migrate to the new food in about two to three weeks.

While you are waiting for them to migrate, you can freeze your excess organics or bury them in your compost bin. In two to three weeks you can harvest the castings from the far end of the bin. The majority of your worms will be enjoying the food at the other end of the bin. Be sure that you do not have too many worms for the size of the bin.

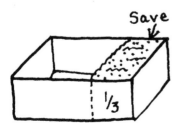

This harvesting method lends itself well to the following technique of *curing* the worm castings.

What is 'curing' castings?

This process has not been discussed very much in the worm world. While you can use your castings as soon as you harvest, often you don't have an immediate need for them. Curing is a way to complete your worm composting efforts, giving you an even more finished product. There are always bits of unfinished matter in the castings at harvest. During the curing, the few worms, hatchlings and cocoons in the pail will continue to eat and finish off the food. It is a simple process and we think you will be pleased with the results.

1) After harvesting, fill a pail or other container with castings.
2) Slowly add water that has been set out for 24 hours until the castings are moist to the touch, this may take one quart or less.
3) Cover with a lid and put it in a shady place for two weeks to a month.
4) Check for the moist-to-the-touch test after two weeks, adding water if dry. Remember to leave the water out for 24 hours to allow the chlorine and other gases to escape into the air before adding it to the curing container.

Curing the castings makes them easier to work with when you are making your potting mixtures. When you open the pail, the castings will have a nice even consistency and you should have more worms!

Using castings

There are many uses for worm castings in gardening. The castings will have a nice even consistency and you should have more Depending on the worms' diet, castings can contain up to eleven times the amount of nitrogen, potassium and phosphorous available in soil unworked by worms.[11] Castings also have a favorable pH, ranging from 5.5 to 7.1 making them a safe, non-burning starter food for seedlings.

---

[11]Edwards, C. A. & Bohlen, P. J.; *The Biology and Ecology of Earthworms*, page 184.

Use castings to enrich potting soil. The sheathing property of the worm castings causes them to clump together. Add about one-quarter castings to three-quarters of your regular soil mix. If you exceed this proportion, the potting soil will become hard-packed with cement-like consistency.

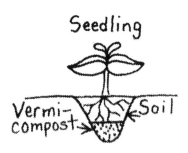

Seeds can be started in pots or trays with a mixture of castings and your favorite potting soil.

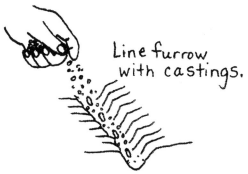

For transplanting seedlings: line the bottom of a row with castings to a depth of two inches, working the castings into the soil before planting. After planting you will see positive results in your seedlings, with little, if any transplant shock.

To make a wonderful liquid fertilizer, often referred to as *worm tea*, tie one cup of castings in an old sock or stocking. Soak in one gallon of water overnight or a few days and use as you would a commercial brand of fertilizer. It can be stored up to two months in a capped container. Use as you like on your house plants. Worm tea also can be used as a foliar spray on both your indoor and outdoor plants.

Mature shrubs, ornamental and fruit trees all benefit from worm castings. Put the castings two to three inches from the base of the plant and in a circular pattern continue out to the drip line. Nutrients in the castings are water soluble, so watering and working the castings into the soil immediately after application will result in healthier plants. This fertilizing treatment can be repeated throughout the season; something you would have done with costly manufactured chemicals in the past.

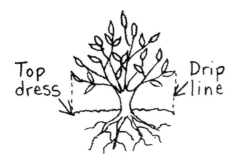

To use castings on your lawn, pre-water the area you will treat. Broadcast the castings just like any commercial fertilizer. Re-water the lawn afterward to begin leaching nutrients into the root zone.

You can now use your worm castings instead of commercial fertilizers. Plus, you have made a herd of worms a very happy home.

Just think, you may never again use your electric garbage disposal and you are saving all the rinse water which would have gone down the drain. Your electric bill is lower. And, you have reduced your dependence on landfills by turning waste into a valuable resource.

Finally, in a small but important way, you have improved the environment. This is a better world because of your efforts.

Now you have completed the cycle! Good job, well done!

Earthworm Myths
Twenty Questions to Check Your Worm Savvy

There are many misconceptions regarding worms. Some are false and some are fanciful. Here is a list of worm myths.

1. If a worm is cut in half, two worms are the result.

2. Worms only eat, mate and sleep at night.

3. Worms are harmful to plants and their roots.

4. Keeping a worm bin is hard work.

5. All worms are the same.

6. There are both male and female worms.

7. Worms carry diseases.

8. Worms need light to see.

9. Worms like dry, hot places.

10. Worms don't make any noise.

11. Worms can't hear.

12. Worms chew their food like cows.

13. Worms have no blood.

14. Worms are only good as fish bait.

15. Worms live just one year in a worm bin.

16. Worms can only move forward.

17. Worms mate for life.

18. Worm bins stink.

19. Worms are vegetarians.

20. Worms won't eat dairy products or meat.

Answers:

1. *False.* If a worm is cut in half it will die. On rare occasions a worm which loses the last one-quarter of it's body under the best of circumstances will survive. The severed piece will not.

2. *False.* If the worm's environment remains dark they will keep active throughout the daylight hours.

3. *False.* The worms are the plants' biggest supporters.

4. *True/False.* Life is hard work, but the benefits of keeping a worm bin far outweigh the efforts.

5. *False.* There are over 3,000 different species of worms that are as diverse in appearance and living environments as humans.

6. *True.* But most species of worms are hermaphroditic.

7. *True.* All creatures are susceptible to diseases, however in many cases earthworms are able to reverse pathogenic effects in their environment and there are no worm diseases communicable to humans. It is important to wear gloves while working in the worm bin to protect worm cocoons from human body oils.

8. *False.* Worms have no eyes, but they do sense light and dive for cover.

9. *False.* Worms for the most part need a cool, moist place to live.

10. *False.* At the peak of their breeding season you may hear worm song in the bin. This sound is created by them sliding by each other and feeding in large numbers.

11. *True/False.* Worms don't have ears, but are adversely affected by many vibrations, such as car travel or being placed too close to the clothes dryer or stereo.

12. *False.* Worms don't chew, they have no teeth. Worms have a crop and a gizzard for grinding their food.

13. *False.* Worms have hemoglobin and a circulatory system much like ours.

14. *False.* Ecological sportsmen use artificial lures. Worms are 76% protein and used as human food in some parts of the world.

15. *False.* A research team at the University of Illinois kept one happy worm alive for 15 years.

16. *False.* Worms use their setae and longitudinal muscles to move in any direction necessary.

17. *False.* Worms aren't picky, any mate will do.

18. *False.* A mismanaged bin will create objectionable odor, but a well managed, well aerated bin has the sweet fragrance of the forest floor.

19. *False.* Worms will eat anything organic. If you died in the forest, the worms eventually would eat everything but the fillings in your teeth.

20. *False.* Worms will eat animal matter, but if not handled properly these items can attract rodents and flies. We recommend you either do not include these in your worm bin, or bury this waste in the bin to avoid the problems.

Appendix

Paper vs. Plastic in the Waste Stream

Common Organic Waste Resources

Information and Resources for Worm Related Items

Glossary

Further Reading

Plastic vs. Paper in the Waste Stream

Dr. Jan Beyea, Senior Policy Scientist, National Audubon Society, has said, "The brown paper bags used in most supermarkets are made from virgin paper, without contributions from recycled materials. Paper making pollutes the water, releases dioxin, contributes to acid rain and costs trees...Heretical as it may sound, some uses of virgin paper can be more damaging to wildlife than plastic substitutes. . .If they (consumers) support virgin paper bags over plastic bags, they are implicitly supporting higher levels of pollution,"

Let's begin with a stack of 2,000 paper and 2,000 plastic bags. Paper bags would stack 7.6 feet tall compared to 7.25 inches for the plastic bags. Transporting the paper bags requires the use of trucks, their maintenance and pollution, to reach the user destination. Plastic bags are also delivered by truck, but, by volume, many more can be delivered per truck compared to the paper bags.

Paper bags are recyclable, a natural fiber which can be composted. True, but how many people actually follow through this cycle? Due to processing advancements, plastic grocery bags are now recyclable at 14,000 sites nationwide.

A Few Second-Use Suggestions

Paper Bags
- ✓ Use paper bags, if you prefer them. They can be very utilitarian.
- ✓ Give them to children to draw on; make hats and masks; puppets and rocket ships. Use your imagination and your child's .
- ✓ Paper bags make sturdy book covers, drawer liners and packing material.
- ✓ When their usefulness has ended, soak them and place them in the worm or compost bin.

Plastic Bags
✓ Recycle plastic into the garden as a weed retardant in place of plastic film.
✓ Knot and tie plastic bags together to make a garden trellis
✓ Support trellised fruit and vegetables, this also helps prevent sun-scorch and bird damage.
✓ Use plastic bags as packing material

Years could be spent studying the plastic versus paper issue, but this has already been done at great length by the Plastic Bag Information Clearinghouse. Their address is in the Information and Resource List, at the end of the text.

Another approach to this issue is to purchase or make cloth bags to carry to the market and on other shopping excursions, a common practice in Europe. These bags are easily portable, strong, washable and last for years.

Common Organic Waste Resources

This list is provided to open your mind to the vast array of sources for organic materials. Not all of these organic resources are appropriate for the worm bin. We hope you recycle and compost, as well.

☐ Food — vegetable and animal products, trimmings and scraps. Coffee grounds - nearly every home has these, and tea bags. Most restaurants will give you coffee grounds if you provide a container and agree to pick them up every few days.

☐ Hair — very high in nitrogen. Barber shops and stylists have an ample supply.

☐ Clothing — wool, cotton, leather, rayon, silk, down, and feathers.

☐ Vacuum cleaner bags — what is vacuumed and swept up from floor surfaces is mostly organic, dirt, hair, food, rug fibers.

☐ Paper products — junk mail, newsprint, all paper rolls for home use, all cardboard-like stiff magazine inserts, paper, tissues, paper towels. Avoid plastics, plasticine and carbon paper.

☐ Animal waste — manure, blood, bones and hair.

☐ Leaves — from your own yard and garden, or they can be found bagged and ready to go on trash day at the curb, from your neighbors.

☐ Seaweed — often washed up on beaches. It is a good source of trace elements, but rinse off the salt before composting or adding to the worm bin.

☐ Wood products — lumber, sawdust (this takes a great deal of nitrogen to breakdown).

☐ Yard wastes — weeds, spent flowers, old plants, grass clippings and prunings.

☐ Tree trimmings — local utilities and maintenance departments trim the trees to keep the wires and utilities safe. They also chip and shred them. Call and ask if they will deliver the chips. You can use them on garden paths and also for your composting. The loads are very large, so be sure you have room for a huge pile!

We are confident you will be creative and think of many other items that can be taken out of the waste stream to reduce the amount of material going to landfills. As Master Composters, we feel the effort expended is well worth the time you give to recycle, reuse and renew our planet!

Information and Resources
for Worm Related Items

Mail Order Catalogs

Smith and Hawkins: worm supplies, information, books, worms.
1-800-776-3336

Park Seed: worm related supplies and worms.
1-800-845-3369

Worm's Way: worm supplies and worms.
1-800-274-9676

Gardeners Supply Company: for worm related items, worms and peat free blocks.
1-800-955-3370

Information

University Agricultural Cooperative Extension Office, affiliated with the U.S. Department of Agriculture and local agricultural offices: check your local phone listings.

Biocycle, Journal of Composting and Recycling, monthly magazine. Topical, cutting edge in composting and recycling.
610-967-4135

Earthworm Buyer's Guide: A Directory of Earthworm Hatcheries in the USA and Canada, published biannually, Shields Publications, P. O. Box 669, Eagle River, WI 54521

Plastic Bag Information Clearing House
1-800-438-5856, e-mail: pbaino@aol.com

Worm Digest, quarterly international newspaper. Carries the latest developments in the worm world, both large and small projects;

home, school, industrial; state and national implications and applications of worms. Ads for worms, related products and networking for worm enthusiasts. Edible City Resource Center, P.O. Box 544, Eugene, OR 97440-0544

e-mail: mail@wormdigest.org

web site: http://www.wormdigest.org

Glossary

actinomycete - a group of unicellular microorganisms, with characteristics of both bacteria and molds. They decompose dead plants and are responsible for the rich, earthy smell of compost and vermicompost.

aeration - exposure to air, allowing for an exchange of gases, related to porosity in soil.

aggregate - a mass of distinct things that when gathered together adhere to each other and make up a whole or a total.

alkali - any of various base chemicals such as hydroxides and carbonates, which neutralize acids and form salts.

anaerobic - adjective describing an organism that can live or function in the absence of oxygen. This condition often produces foul odors.

anterior - positioned in or toward the front, opposite of posterior.

bedding - a moisture retaining medium used for worms to live in, as in a worm bin.

buccal cavity - the mouth cavity

bulbous - shaped like or having a bulb.

calciferous glands - glands within the worm mouth that produce a calcium carbonate substance that enables the worm to expel 'sheathed castings.'

castings - see worm castings.

cerebral ganglion - a collection of nerve cells that serve as the brain for a worm.

chitin - a tough covering secreted by the epidermis and forming the outer wall of the worm body.

clitellum - an organ present in sexually mature worms. Noticeable as a swollen region on a worm, slightly lighter in color, containing gland cells which secrete the cocoon material.

cocoon - a protective envelope encompassing embryonic worms until they are develop enough to hatch.

coelomic fluid - a fluid that is expelled through the anus of the worm in times of stress.

composting - the biological degradation or breakdown of organic matter by a managed process.

crop - digestive organ in the anterior part of a worm consisting of a thin-walled sac just behind the esophagus

dorsal - on the under side of a worm.

dorsal blood vessel - the blood vessel on the under side of the worm.

earthworm - descriptive term for segmented worms in the phylum Annelida. All earthworms have no skeleton, are headless, eyeless, toothless and without antennae.

egg case - see cocoon.

Eisenia fetida - scientific name for red worm or red wiggler.

emulsion - an oily substance held in suspension in a watery liquid by means of a gummy substance.

enchytraeid - white worm one-quarter to one inch long. Similar in appearance to baby red worms, except they have no hemoglobin pigment and even baby red worms are slightly pink. They eat organic material and are fine in the worm bin.

esophagus - a tube through which food passes from the pharynx to the crop, then gizzard of the worm.

excrete - to separate and eliminate from an organic body.

exoskeleton - the hard external supporting outer body, as in most insects, as in a beetle.

foliar spray - a liquid (fertilizer) that when sprayed on a plant can be absorbed and used by the plant.

fungi - plural of fungus, simple organisms that lack photosynthetic pigment.

gestation - that span of time from fertilization and egg formation to worm hatchlings emerging from the cocoon.

girdle - see clitellum.

gizzard - a digestive organ in the anterior of the worm, its muscular contractions help grind food.

hemoglobin - iron-carrying compound in the blood which carries oxygen to the tissues.

humus - a dark organic material in soils produced by decomposition of organic matter, essential to soil fertility.

leaching - the filtering of a substance through soil by water or using water to wash out chemical content.

loam - a mixture of clay, sand and organic matter that produces a rich soil.

Lumbricus rubellus - scientific name for a red worm.

Lumbricus terrestris - scientific name for large burrowing night crawling worms.

mold - a furry growth on moist decaying organic material, usually a fungus.

mucus - a slimy secretion of the mucus membranes that coats and protects, as in the lining of the lungs.

nephridia - pertaining to the urinary excretion of the worm.

night crawler - see *Lumbricus terrestris*. A worm not used for vermicomposting.

nutrient - a life sustaining substance, a food source.

ovum - a mature, female cell, after fertilization, which can develop into a member of the same species.

parthenogenic - the reproduction and the development of a species by an unfertilized ovum, as in some species of worms.

pathogen - disease producing organism often occurring in waste materials.

peat - partially decomposed remains of plants used as a soil amendment, fuel, mulch and potting soil. Has great moisture-retaining abilities within the worm bin.

peat-free - without peat, a coconut fiber similar to and substitute for regular peat but not decomposed. Will expand up to ten times its original mass when placed in water.

percolate - a liquid passing through a porous space.

pH - a measure of concentration of hydrogen ions in a solution. Scale ranges from 0 to 14. A pH of 7 is considered neutral: below 7, acid; above 7, alkaline.

pharynx - muscular region of the digestive tract immediately behind a worm's mouth.

population density - specific number of worms in a given area, usually designated in worms per square foot.

porosity - permeability by water.

potting soil - a medium for growing plants.

prostomium - a fleshy lobe-shaped protrusion above the mouth.
protein - complex molecule containing carbon, oxygen, hydrogen and nitrogen, a major constituent of meat. Worms are approximately 76% protein.
protozoa - microscopic animals.
putrification - anaerobic (without oxygen) decomposition of organic material, associated with unsavory smells, like rotting eggs.
red worm - common name for several worms with red color including, *Eisena fetida* and *Lumbricus rubellus*. *E. fetida* is the chosen worm for vermicomposting.
vermicomposting - composting with worms.
segmented - any part of a body that is separated by another part, as in the circular rings of the worm.
setae - short stiff hairs or bristles. Four pairs occur on most of the worm's segments helping the worm to move through soil.
side-dressing - application of nutrients on the soil surface away from the stems of plants.
slough - to glide or slip off, as in a cocoon sliding over the worm head.
soil conditioner - any of various organic or inorganic substances added to the soil to improve it. May affect resistance to erosion, permeability to air and water, texture, resistance to surface crusting.
soil amendment - (see soil conditioner) - any organic or inorganic substance, such as lime, sulfur or sawdust, used to alter the properties of the soil. Fertilizers are one type, however, others, like soil conditioners, do not have any nutritive value.
species - basic category of biological classification among organisms with common characteristics, characterized by those which can breed together but cannot breed with members of other species.
surface crusting - soil texture at the top affecting ease of cultivation.
ventral - pertaining to the top of the worm body.
vermi - borrowed from Latin meaning "worm", used in the formation of compound words.

vermicompost - a mixture of partially decomposed organic material, bedding, worms, cocoons and other bin wildlife from the worm bin; compost from worms.

vermicomposting - a process by which worms changes organic resources into worm castings through digestion of organic matter.

vermiculture - raising worms in controlled spaces such as bins, pits or windrows, using controlled conditions.

white worms - see enchytraeid.

worm casting - the dark, granular, fertile excrement of worms rich in plant nutrients. Worm manure.

worm tea - a liquid fertilizer made from worm castings, which can be used on indoor and outdoor plants, trees and shrubs.

Further Reading

Barrett, Thomas, J.,1976: *Harnessing The Earthworm,* Shields Publications, WI

Bhawalker, Uday, 1993, *Turning Garbage Into Gold,* The Studio, Bhawalker Earthworm Institute, Pune, India, 411 037

Can-O-Worms ™ Manual, (a how-to book for a tiered round bin,) 1995. Made in Australia, patent pending in the US # 29/023712. Can-O-Worms™ is the trademark of N. Nattrass

Carr, Ann; Smith, Miranda; Gilkeson, Linda A.; Smillie, Joseph; Wolf, Bill; Bradley, Fern Marshall, 1991: *Rodale's Chemical-Free Yard and Garden,* Random Press, New York, NY

Crowe, Mary B. and Bowen, Gladys S. 1970: *With Tails We Win!* Shields Publications, WI

Earthworm Buyer's Guide: A Directory of Earthworm Hatcheries in the USA and Canada, Shields Publications, Eagle River, WI

Edwards, C. A. and Bohlen, P. J. 1995: *Biology and Ecology of Earthworms,* Chapman Hall, New York, NY

Gaddie, Ronald E. Sr. & Douglas, E. Donald, 1975: *Earthworms For Ecology and Profit,* Bookworm Publications, CA

Holwager, George H., 1952: *The Production and Sale of Larger Redworms,* Shields Publications, WI

Hopp, Dr. Henry, 1978: *What Every Gardener Should Know About Earthworms,* Storey Communications, Storey, VT

Martin, Deborah L. & Gershuny, Grace; Editors, 1992, *The Rodale Book of Composting,* Publisher: Rodale Press, PA

Morgan, Charlie, 1961: *Earthworm Feeds and Feeding*, Shields Publications, WI

Morgan, Charlie, 1978: *The Worm Farm, A Diary*, Shields Publications, WI

Shields, Earl B., 1959: *Raising Earthworms For Profit*, Shields Publications, WI

USDA and University of California at Berkeley, 1977: *Earthworm Biology and Reproduction*, Leaflet 2828, UC Division of Agricultural Science, Berkeley, CA

Periodicals

Ag-Sieve, quarterly newsletter, information packet #1, 1996. Landeck, Jonathan Publications, PA Tel:610-683-1400 http://www.envirolink.org/pubs/index.html or

Biocycle, Journal of Composting and Recycling, monthly magazine, 1960 to present. Goldstein, Jerome, Editor. The JG Press, Inc. Tel:(610) 967-4135

Earthworm Buyer's Guide: A Directory of Earthworm Hatcheries in the USA and Canada, published biannually, Shields Publications, P. O. Box 669, Eagle River, WI 54521

Worm Digest, quarterly newspaper, 1993 to present. White, Stephan, Associate Editor; S. Zorba, Managing Editor; with contributing editors. Edible City Resource Center, P.O. Box 544, Eugene, OR 97440-0544

Books for Children

Appelhof, Mary, 1979: *Worms Eat My Garbage*, Flower Press, Kalamazoo, MI

Appelhof, Mary & Fenton, Mary Francis, 1993: *Worms Eat Our Garbage*, Flower Press, Kalamazoo, MI

Darling, Louis & Lois, 1972: *Worms,* William Morrow and Company, New York, NY

Hess, Lilo, 1979: *The Amazing Earthworm*, Charles Scribner's Sons, New York, NY

Hogner, Dorothy Childs, 1953: *Earthworms*, Thomas Crowell, New York, NY

Jennings, Terry, 1988: *Earthworms,* Gloucester Press, New York, NY

Lauber, Patricia, 1973: *Earthworms, Underground Farmers*, self published, USA

O'Hagan, Caroline, 1980: *It's Easy To Have A Worm Visit You,* Lothrop, Lee and Shepard Books, New York, NY

Pringle, Laurence, 1973: *Twist, Wiggle and Squirm*, Thomas Crowell, New York, NY

Simon, Seymour, 1969: *Discovering Earthworms*, McGraw Hill, New York, NY

Index

actinomycetes 10, 23

aggregate 7, 82

alimentary canal 23

ambient 30, 34

anecic 14

annelid 12, 13, 19

antibiotic 23

anus 23, 82

bacteria 10, 17, 23, 25, 40, 46, 50, 51, 82

buccal cavity 23, 82

burrow 9, 11, 14, 15

burrowing 13, 15, 83

calcium 19, 28, 29, 82

carbon dioxide 25, 61

castings 4, 9-11, 15, 23, 35, 36, 40, 43, 56, 61, 62, 65-68,
82, 85

cerebral ganglion 19, 21, 82

chitin 21, 82

chlorine 39, 42, 66

clitellum 28, 31, 82, 83

cocoon 28-31, 49, 61, 62, 66, 71, 82, 83, 85

coelomic fluid 17, 82

composting 2-5, 10, 11, 15, 17, 18, 33, 39, 43, 46, 49-52,
54, 60, 65, 66,78, 80, 82, 84, 85, 87, 88

consume 14, 15, 19, 37, 38, 51

consumption 17, 51

crop 23, 25, 40, 71, 82, 83

cuticle 19, 21

decomposition 4, 7, 17, 38, 61, 84

dorsal 21, 82, 83

ecosystem 42

Eisenia fetida 15, 17-19, 27, 29-31, 36, 43, 83, 84

endogeic 14

enzyme 23
epidermis 19, 82
esophagus 23, 82, 83
exoskeleton 21, 83
fertilizer 2, 5, 10, 46, 66-68, 83, 85
foliar 67, 83
fungi 10, 23, 25, 56, 83
ganglion 19, 21, 82
gestation 30, 83
gizzard 23, 25, 40, 62, 71, 83
harvest 29, 34, 36, 41-43, 49, 54, 56, 61, 62, 65, 66
hatchlings 28-30, 66, 83
hemoglobin 30, 72, 83
hermaphroditic 13, 27, 71
humus 5-7, 9, 83
incubation 29
inorganic 2, 5-7, 85
insecticide 23
intestine 10, 19, 23
landfill 2-5, 68, 79
loam 7, 83
Lumbricus 15, 83, 84
macrobiotic 25
macroorganisms 56
microorganisms 5, 10, 25, 56, 82
mold 10, 25, 56, 82, 83
molecular 38
mucus 10, 23, 28, 30, 84
nitrogen 10, 66, 78, 84
nocturnal 19
nutrient . . . 2, 4-7, 9-11, 15, 21, 23, 30, 38, 46, 61, 66-68, 84, 85
organic 2-7, 9-15, 17, 18, 23, 25, 28, 32, 33, 37, 38, 40, 43-
                                            45, 50, 60, 65, 72, 74, 78, 82-85
ovum 28, 84
oxygen 18, 21, 25, 38, 61, 82-84
parthenogenic 27, 84
particle 6, 7, 9, 10, 15, 23, 40

pathogenic 46, 71
peat 32, 41, 80, 84
percolate 7, 84
pH 19, 46, 50, 66, 84
pharynx 23, 83, 84
phosphorous 10, 66
porous 6, 7, 39, 84
potassium 10, 66
prostomium 17, 84
protozoa 23, 84
recycling 3, 9, 12, 35, 50, 56, 59, 76, 78-80, 88
red wiggler 17, 18, 27, 29, 31, 32, 39, 83
red worm 15, 83, 84
segment 21
segmented 13, 83, 85
seminal fluid 28
setae 21, 28, 72, 85
soil 2-7, 9-11, 13-15, 17-19, 21, 23, 32, 33, 40, 43, 46, 47,
53, 59, 61, 66, 67, 82-85
soil amendment 2, 3, 5, 11, 15, 61, 84, 85
terrestrial 13, 18
top feeder 14, 15, 17-19, 36
ventral 21, 85
vermicompost . 3-5, 15, 17, 22, 33, 46, 50, 60, 62, 65, 82, 84, 85
waste stream 3, 4, 32, 74, 76, 79
worm bin 10, 19, 34, 36-39, 42, 44-46, 54, 55, 59-61, 70-
72, 78, 82-85
worm castings 4, 10, 11, 15, 35, 61, 62, 65-68, 82, 85
worm compost 10, 15, 54
worm tea 38, 67, 85

Weekly Worm Feeding Record

Monday:

Tuesday:

Wednesday:

Thursday:

Friday:

Saturday:

Sunday:

Monthly Worm Notes

Week One:

Week Two:

Week Three:

Week Four:

Week Five:

Quarterly Worm Harvest Log

First Quarter:
    Date:
    Pounds of Worms:
    Pounds of Castings:

Second Quarter:
    Date:
    Pounds of Worms:
    Pounds of Castings:

Third Quarter:
    Date:
    Pounds of Worms:
    Pounds of Castings:

Fourth Quarter:
    Date:
    Pounds of Worms:
    Pounds of Castings:

Worm Notes

Worm Notes

Worm Notes

Worm Notes